W9-BYT-107

Green Products by Design

Choices for a Cleaner Environment

CONGRESS OF THE UNITED STATES
OFFICE OF TECHNOLOGY ASSESSMENT

Cover credit: The design depicted on the cover is a blueprint of the NCR 7731 Personal Image Processor. This optical imaging device incorporates a number of green design concepts: modular components, parts consolidation, design for disassembly, and the use of recycled materials.
SOURCE: NCR Canada, Ltd.

This report is printed on paper that is 100 percent recycled (15 percent post-consumer material).

Recommended Citation:

U.S. Congress, Office of Technology Assessment, *Green Products by Design: Choices for a Cleaner Environment*, OTA-E-541 (Washington, DC: U.S. Government Printing Office, October 1992).

For sale by the U.S. Government Printing Office
Superintendent of Documents, Mail Stop: SSOP, Washington, DC 20402-9328
ISBN 0-16-038066-9

Foreword

As the United States and other nations begin to get serious about an array of potentially significant environmental threats, from global climate change to local groundwater contamination, traditional formulas of environmental management are being reassessed. The remediation or ''end-of-pipe'' strategies of the past 20 years are unlikely to provide satisfactory, cost-effective protection of ecosystems and human health in the future. Systematic change is needed.

Increasingly, product design is being viewed as a possible catalyst in transforming societal patterns of production and consumption. Product design is an important environmental focal point, because design decisions directly and indirectly determine levels of resource use and the composition of waste streams. By placing a greater emphasis on design, environmental problems can be addressed in a proactive manner.

In this report, OTA provides a conceptual overview of how designers might integrate environmental concerns with traditional design objectives, and how policymakers can best take advantage of such opportunities. Although the concept of ''green'' design is gathering momentum, a number of technical, behavioral, economic, and informational barriers need to be addressed. By relying solely on existing policies and industrial practices, the full potential of green design will not be realized.

Because product design encompasses the most crucial decisionmaking activities of companies, the consideration of environmental objectives by designers could have important competitive implications. Market opportunities for environmentally sensitive goods and services are expanding. Examples of ''green'' products and ''clean'' technologies are beginning to appear across a wide spectrum of industries.

This study, requested by the House Committees on Science, Space, and Technology, and on Energy and Commerce, builds on previous OTA work dealing with U.S. hazardous waste and municipal solid waste policy. These previous studies acknowledged the importance of product design as a tool for reducing wastes and managing materials, but did not explore the idea in detail.

OTA appreciates the assistance provided by its contractors and the advisory panel, as well as the many reviewers whose comments helped to ensure the accuracy of the report.

JOHN H. GIBBONS
Director

Advisory Panel—Green Products by Design:
Choices for a Cleaner Environment

Indira Nair, *Chairman*
Carnegie Mellon University

Frank van den Akker
Ministry of Housing, Physical Planning,
    and Environment

Harvey Alter
U.S. Chamber of Commerce

David Chittick
AT&T

Michael DeCata
GE Plastics

Michael A. Gallo
Robert Wood Johnson Medical School

Robert Garino
Institute of Scrap Recycling Industries

Gil Gavlin
Gavlin Associates, Inc.

Denis Hayes
Green Seal

Kenneth Hunnibell
Rhode Island School of Design

Richard L. Klimisch
General Motors Corp.

Michael H. Levin
Nixon, Hargrave, Devans & Doyle

Stewart Mosberg
Walter Dorwin Teague Associates, Inc.

Thomas Rattray
Procter and Gamble Co.

Cliff Russell
Vanderbilt Institute for Public Policy Studies

Mary T. Sheil
New Jersey Department of Environmental
    Protection and Energy

T.S. Sudarshan
Materials Modifications, Inc.

Don Theissen
3M Co.

Jeanne Wirka
Consultant

Dennis A. Yao
Federal Trade Commission

NOTE: OTA appreciates and is grateful for the valuable assistance and thoughtful critiques provided by the advisory panel members. The panel does not, however, necessarily approve, disapprove, or endorse this report. OTA assumes full responsibility for the report and the accuracy of its contents.

OTA Project Staff—Green Products by Design: Choices for a Cleaner Environment

Lionel S. Johns, *Assistant Director, OTA*
Energy, Materials, and International Security Division

Peter D. Blair, *Energy and Materials Program Manager*

Project Staff

Gregory Eyring, *Project Director*

Matthew Weinberg, *Analyst*

David Jensen, *Analyst*

Joe Raguso, *Contractor*

Administrative Staff

Lillian Chapman, *Office Administrator*

Linda Long, *Administrative Secretary*

Tina Aikens, *Secretary*

Contractors

Jill Watz and Thomas Nunno
ChemCycle Corp.

Frank Field
FieldWorks

William Franklin and Robert Hunt
Franklin Associates

John Kusz
Independent Consultant

Robert Kerr and Lisa Lambrecht
Kerr and Associates, Inc.

Jill Shankleman and David Festa
Environmental Resources Limited

Outside Reviewers

Frank Ackerman
Tellus Institute

Braden Allenby
National Academy of Engineering

John Atcheson
U.S. Environmental Protection Agency

Jesse Ausubel
Rockefeller University

Robert Ayres
Carnegie Mellon University

Charles Burnette
Philadelphia College of Art and Design

Richard Denison
Environmental Defense Fund

Tom Donnelly
Council for Solid Waste Solutions

Pat Eagan
University of Wisconsin at Madison

John Ehrenfeld
Massachusetts Institute of Technology

Peter Eisenberger
Princeton University

Michael Fischer
Society of the Plastics Industries

Walter J. Foley
Coalition of Northeastern Governors

Ken Geiser
University of Massachusetts

Eun-Sook Goidel
U.S. Environmental Protection Agency

Paul Kaldjian
Montgomery Department of Environmental
    Protection

Greg Keoleian
University of Michigan

David Lennert
Procter and Gamble Co.

Howard Levenson
California Integrated Waste Management Board

Mark Mazur
Joint Tax Committee

James McCarthy
Congressional Research Service

D. Navin chandra
Carnegie Mellon University

Doug Nutter
GE Plastics

Winfred Phillips
University of Florida

David Ruller
City of Alexandria

Walter Shaub
Solid Waste Association of North America

Ann Speicher
American Society for Engineering Education

Ed Stana
Council on Plastics and Packaging in the
    Environment

Arthur J. Zadrozny
ARCO Chemical Co.

Contents

Chapter 1

Executive Summary

Contents

Box

Figures

Tables

Executive Summary

Policymakers should be concerned with product design for two reasons. One is to improve U.S. industrial competitiveness. A strong domestic design capability can slash product development time, improve quality, and reduce the cost of U.S. products. The National Research Council has estimated that 70 percent or more of the costs of product development, manufacture, and use is determined during the initial design stages.[1] Thus, design is a critical determinant of a manufacturer's competitiveness.

The second reason is that product design is a unique point of leverage from which to address environmental problems. Design is the stage where decisions are made regarding the types of resources and manufacturing processes to be used, and these decisions ultimately determine the characteristics of waste streams.[2] By giving designers incentives to consider the environmental impacts of their choices, policymakers can address environmental problems that arise throughout the product life cycle, from the extraction of raw materials to final disposal.[3]

The two design goals of enhancing competitiveness and protecting environmental quality can be consistent. Design strategies that reduce production costs and improve quality often have the benefit of generating less waste and pollution. Moreover, many companies are already using the environmental attributes of their products in their marketing strategies, and polls suggest that consumer demand for "green" products is likely to grow.[4] Many observers believe that those companies that are able to design high-quality, environmentally sound products will enjoy a competitive advantage in the 1990s and beyond.

In a recent review, the National Research Council found that the quality of U.S. engineering design is generally poor, and recommended that the Federal Government make engineering design a national priority to improve competitiveness.[5] In the present study, the Office of Technology Assessment (OTA) finds that better product design offers new opportunities to address environmental problems, but that current government regulations and market practices are not sufficient to fully exploit these opportunities. Therefore, integrating an environmental component into policies to improve U.S. design capabilities is an important policy objective. But policymakers should be careful in how they attempt to achieve this objective. Inappropriate regulation of the environmental attributes of products could perversely lead to more wastes being generated, and could also adversely affect competitiveness.

These findings are particularly relevant in the light of congressional debate concerning the reauthorization of the Resource Conservation and Recovery Act (RCRA) of 1976, the major Federal statute concerning solid waste. The reauthorization debate involves many issues that could affect the design of products, including mandatory recycled content, reduced toxic chemical content, government procurement of recycled products, and environmental labeling, as well as controls on products that cause special waste management problems such as automobile batteries, used oil, and tires.

In reauthorizing RCRA and other environmental laws, Congress can influence product design in two ways. First, it can enact additional environmental regulations—for example, requiring that manufacturers incorporate recycled materials into new products or take back discarded products from consumers. Second, Congress can move toward a strategy of harnessing market forces to encourage manufacturers to make environmentally sound decisions—for example, instituting a fee-rebate system based on the

---

[1] National Research Council, *Improving Engineering Design: Designing for Competitive Advantage* (Washington, DC: National Academy Press, 1991).

[2] While this report focuses primarily on product design rather than process design, it should be recognized that the two are closely related. Many of the research needs and incentives discussed here for product design also apply to process design.

[3] As used here, the term "designers" refers to all decisionmakers who participate in the early stages of product development. This includes a wide variety of disciplines: industrial designers, engineering designers, manufacturing engineers, graphic and packaging designers, as well as managers and marketing professionals.

[4] Green products are those whose manufacture, use, and disposal place a reduced burden on the environment.

[5] National Research Council, op. cit., footnote 1.

Figure 1-1—Stages of the Product Life Cycle

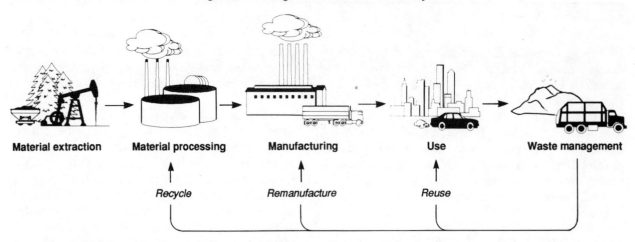

Environmental impacts occur at all stages of a product's life cycle. Design can be employed to reduce these impacts by changing the amount and type of materials used in the product, by creating more efficient manufacturing operations, by reducing the energy and materials consumed during use, and by improving recovery of energy and materials during waste management.

SOURCE: Adapted from D. Navin chandra, The Robotics Institute, Carnegie Mellon University, personal communication, March 1992.

energy efficiency of products, or taxing the industrial emissions of certain toxic chemicals.

Each approach has advantages and disadvantages: regulations can produce swift and predictable results, but they can also impose unnecessary costs on industry and stifle environmentally innovative designs. Economic incentives can provide flexibility, but they can be expensive to administer and are often politically unpopular. The challenge for Congress is to employ a mixture of regulations and economic instruments to give designers the incentives to make choices that promote RCRA's goals of protecting human health and the environment.

PRODUCT DESIGN AND THE ENVIRONMENT

Products affect the environment at many points in their life cycle (figure 1-1). The most visible impact is municipal solid waste (MSW). The trash generated by U.S. households and commercial establishments averages about 4 pounds per person each day. In 1988, the United States generated some 180 million tons of MSW. Landfills in many States are reaching their permitted capacity, and there is increasing public opposition to siting new waste management facilities.

About one-third of MSW by weight consists of product packaging, which has become a major target of environmental policies around the world. Better packaging design can reduce the quantity of this waste significantly. For example, at a recent conference, packaging designers concluded that—given the commitment of top management—new designs could reduce the weight of packaging by an average of 10 percent in 1 year. This would mean a 3 percent drop in MSW from this source alone.

Less visible but potentially more serious environmental impacts occur during raw material extraction, material processing, and product manufacturing. U.S. industry generates some 700 million tons of "hazardous waste" and some 11 *billion* tons of "non-hazardous" solid waste (figure 1-2a).[6] Although the weight of industrial and municipal solid waste cannot be compared directly,[7] these production wastes clearly dwarf municipal solid wastes in their quantity and environmental impact (see figure 1-2b). Product design decisions can have a direct

---

[6] The terms *hazardous* and *non-hazardous* are defined by RCRA subtitles C and D, respectively. See U.S. Congress, Office of Technology Assessment, *Managing Industrial Solid Wastes From Manufacturing, Mining, Oil and Gas Production, and Utility Coal Combustion*, OTA-BP-O-82 (Washington, DC: U.S. Government Printing Office, February 1992).

[7] Up to 70 percent of the weight of industrial solid waste (which includes mining, oil and gas, and manufacturing wastes) consists of wastewater contained in sludges and aqueous solutions.

Photo credit: Environmental Protection Agency

Although municipal solid waste is the most visible result of our consumer society (above), industrial waste streams are a much larger and more serious problem (left). Product design decisions have a direct impact on both industrial and postconsumer waste streams.

Figure 1-2—"Solid" Wastes as Defined Under the Resource Conservation and Recovery Act (RCRA)

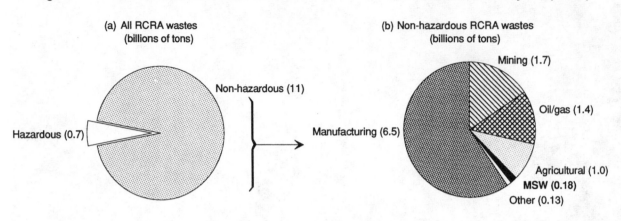

(a) All RCRA wastes
(billions of tons)

Non-hazardous (11)

Hazardous (0.7)

(b) Non-hazardous RCRA wastes
(billions of tons)

Mining (1.7)

Oil/gas (1.4)

Manufacturing (6.5)

Agricultural (1.0)

MSW (0.18)

Other (0.13)

Much of the solid waste produced in the United States is not directly generated by consumers. Municipal solid waste, the focus of much public concern, represents less than 2 percent of all solid waste regulated under RCRA. In contrast, industrial activities produce about 700 million tons of hazardous waste (a) and about 11 *billion* tons of non-hazardous wastes (b).

NOTE: All numbers are estimates. The non-hazardous waste total has been rounded to reflect uncertainty. Much of the "solid" waste defined under RCRA, perhaps as much as 70 percent, consists of wastewater. The terms *hazardous* and *non-hazardous* refer to statutory definitions of Subtitles C and D of RCRA, respectively. The mining wastes shown in (b) exclude mineral processing wastes; the oil/gas wastes in (b) exclude produced waters used for enhanced oil recovery; the "other" category in (b) includes wastes from utility coal combustion.

SOURCE: Adapted from U.S. Congress, Office of Technology Assessment, *Managing Industrial Solid Wastes From Manufacturing, Mining, Oil and Gas Production, and Utility Coal Combustion*, OTA-BP-O-82 (Washington, DC: U.S. Government Printing Office, February 1992).

influence on the manufacturing component of these wastes (about 6.5 billion tons).

Finally, some of the most serious environmental impacts may occur during the actual use of the product. This is particularly true of products that are consumed or dissipated during their use, for example, chlorofluorocarbon (CFC) solvents and coolants, fossil fuels, and pesticides. The environmental releases from these dissipative products can be much larger than those from the associated industrial processes. For example, the State of New Jersey collects data on industrial inputs and outputs of hazardous substances. The data indicate that in 1990, 55 to 99 percent of industrial inputs of five toxic heavy metals (mercury, lead, cadmium, chromium, and nickel) was converted into products (i.e., not released as industrial waste), depending on the metal.[8] Product reformulation and substitution for toxic constituents can help to address these problems.

Behind each of these environmental impacts are critical decisions made during product design. The materials used, energy requirements, recyclability, longevity, and many other environmental attributes of products result directly from design decisions.

Once a product moves from the drawing board into production, its environmental attributes are largely fixed; the key, therefore, is to bring environmental concerns into the front end of the design process.

GREEN DESIGN

Product design is a process of synthesis in which product attributes such as cost, performance, manufacturability, safety, and consumer appeal are considered together. In general, products today are designed without regard for their overall impact on the environment. Nevertheless, many health and environmental laws passed by Congress do influence the environmental attributes of products. Some, such as the Clean Air Act, Clean Water Act and Resource Conservation and Recovery Act, do so indirectly, by raising industry's costs of releasing wastes to the air, water, and land. Others, such as the Toxic Substances Control Act and the Federal Insecticide, Fungicide, and Rodenticide Act, control the use of hazardous chemicals and pesticides directly.

Government regulations typically influence the design process by imposing external constraints, for example, compliance by auto designers with Corpo-

[8] Some heavy metals incorporated into products are eventually recycled, but recycling rates vary substantially by material. For example, more than 50 percent of lead is recycled, but nearly all cadmium is released into the environment.

Photo credit: Courtesy of Discover Magazine (July 1990)

Some bacteria can store energy in polymer-bearing granules that can be collected and made into truly biodegradable packaging like these plastic bottles made from Alcaligenes bacteria by ICI, Ltd.

rate Average Fuel Economy (CAFE) standards, and with auto emissions standards under the Clean Air Act. OTA uses the phrase ''green design'' to mean something qualitatively different: a design process in which environmental attributes are treated as *design objectives*, rather than as *constraints*. A key point is that green design incorporates environmental objectives with minimum loss to product performance, useful life, or functionality.

In OTA's formulation, green design involves two general goals: *waste prevention* and *better materials management* (figure 1-3).[9] Waste prevention refers to activities by manufacturers and consumers that avoid the generation of waste in the first place. Examples include using less material to perform the same function (''light-weighting''), or designing durable products so that faulty or obsolete compo-

nents can be readily replaced, thus extending the product's service life. Better materials management refers to activities that allow product components or materials to be recovered and reused in their highest value-added application. Examples include designing products that can be readily disassembled into constituent materials, or using materials that can be recycled together without the need for separation. These goals should be viewed as complementary: while designers may reduce the quantity of resources used and wastes generated, products and waste streams will still exist and have to be managed.

The idea of green design seems simple, but there is no rigid formula or decision hierarchy for implementing it. One reason is that what is ''green'' depends strongly upon context. While some environmental design objectives are sufficiently compel-

---

[9] This formulation first appeared in U.S. Congress, Office of Technology Assessment, *Facing America's Trash: What Next for Municipal Solid Waste*, OTA-O-424 (Washington, DC: U.S. Government Printing Office, October 1989).

Figure 1-3—The Dual Goals of Green Design

Green design consists of two complementary goals. Design for waste prevention avoids the generation of waste in the first place; design for better materials management facilitates the handling of products at the end of their service life.
SOURCE: Office of Technology Assessment, 1992.

ling to apply to many different products (e.g., avoiding the use of CFCs), in general OTA expects that green choices will only become clear with respect to specific classes of products or production networks. What constitutes green design may depend on such factors as: the length of product life; product performance, safety, and reliability; toxicity of constituents and available substitutes; specific waste management technologies; and the local conditions under which the product is used and disposed.

Design Tradeoffs

With technologies available to create new materials and to combine conventional materials in new ways, designers are faced with more choices than ever before. One result is that products are becoming more complex and specialized. For example, a typical laundry detergent now contains over 25 different ingredients.

These choices often involve environmental dilemmas. Tradeoffs may be required, not only between traditional design objectives and environmental objectives, but even among environmental objectives themselves—for example, waste prevention versus recyclability.

As an illustration, consider the cross section of a modern snack chip bag depicted schematically in figure 1-4. The combination of extremely thin layers of several different materials produces a lightweight package that meets a variety of needs (e.g., preserving freshness, indicating tampering, and providing product information). The use of so many materials

effectively inhibits recycling. On the other hand, the package has waste prevention attributes; it is much lighter than an equivalent package made of a single material and provides a longer shelf life, resulting in less food waste. Even this relatively simple product demonstrates the difficulties of measuring green design.

Similar tradeoffs may occur between other attributes, such as energy efficiency and toxicity. For example, energy-efficient, high-temperature superconductors contain a variety of heavy metals, and toxic chemicals are required to manufacture photovoltaic cells. In general, every design will have its own set of environmental pluses and minuses.

Environmental Aspects of Products and Systems

From an environmental point of view, it is simplistic to consider the impact of a product in isolation from the production and consumption systems in which it functions. Is a computer, for instance, a green product? Considered on its own, probably not. The manufacture of a computer requires large volumes of hazardous chemicals and solvents, and heavy metals used in solder, wiring, and display screens are a significant contributor to the heavy metal content of MSW.

But the same computer could be used to increase the efficiency of a manufacturing process, thus avoiding the use of many tons of raw materials and the generation of many tons of wastes. From this perspective, the computer is an enabling technology

Figure 1-4—Cross-Section of a Snack Chip Bag

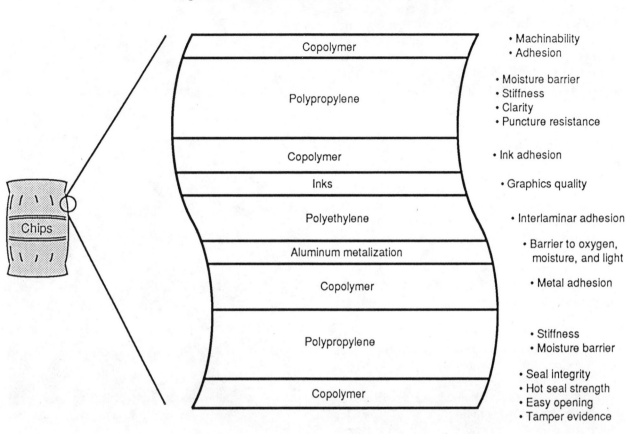

This cross-section of a snack chip bag illustrates the complexity of modern packaging. The bag is approximately 0.002 inches thick, and consists of nine different layers, each with a specific function. While such complexity can inhibit recycling efforts, it also can reduce the overall weight of the bag, and keep food fresher, thus providing waste prevention benefits.

SOURCE: Council on Plastics and Packaging in the Environment.

that reduces the environmental impact of the production system as a whole.

This illustrates an important OTA finding: **green design is likely to have its largest impact in the context of changing the overall systems in which products are manufactured, used, and disposed, rather than in changing the composition of products per se.** For instance, designing lighter fast-food packaging is well and good; but 80 percent of the waste from a typical fast food restaurant is generated *behind the counter*, where consumers never see it. Addressing this larger problem requires that designers establish cooperative relationships with their suppliers and waste management providers in order to manage materials flows in an environmentally sound way.

There may appear to be few incentives for industry to consider such dramatic changes in existing production networks. After all, longstanding relationships among manufacturers and suppliers may have to change, and millions of dollars may be invested in the existing infrastructure for production and distribution. Such changes are not generally within the purview of product designers. Indeed, a systems design approach implies the elevation of the product design function to the level of strategic business planning, and a shift in perception by top management in which environmental quality is viewed not as a cost, but as a strategic business opportunity.

But changes of comparable magnitude are already underway. Many manufacturers are rethinking their business relationships with suppliers and customers in order to implement total quality management and concurrent engineering programs. New government regulations in Europe that give manufacturers responsibility for the environmental fate of their products are also encouraging this approach. For example, Germany's proposed law requiring automakers to take back and recycle automobiles has stimulated the German automobile industry to develop new cooperative strategies for auto design, manufacturing, and recycling (see box 1-A).

Policy Implications

These findings have a number of policy implications:

- The environmental evaluation of a product or design should not be based on a single attribute, such as recyclability. Rather, some balancing of pluses and minuses will be required over the entire life cycle.
- The trend toward increasing product complexity seems certain to make the environmental evaluation of products more difficult and expensive in the future.
- Policies to encourage green design should be flexible enough to accommodate the rapid pace of technological change and the broad array of design choices and tradeoffs.
- The biggest environmental gains will likely come from policies that provide incentives for greener production and consumption systems, not just greener products.

GROWING INTEREST IN GREEN DESIGN

The concept of green design is not new. During the 1970s and 1980s, ideas such as design for remanufacturing and design for recycling were developed in technical journals and conferences. At the time, the concept did not receive much attention from policymakers or the public, but green design has enjoyed a renaissance in the past few years. Rising interest among industry groups and design societies around the world is indicated by the proliferation of books, newsletters, and published papers on the subject.

One area of particular interest is the awarding of "eco-labels" to products that are judged to be environmentally preferred compared with alternative products. Germany, Canada, Japan, the Nordic countries, and the European Community all either have government-funded eco-labeling programs, or will have them in the near future. The United States

Photo credit: Institute of Scrap Recycling Industries, Inc.

Photo credit: GE Plastics

Top: When automobile hulks are recycled, most of the metals are recovered. The nonmetal components (plastic, rubber, fabric, and glass) end up as shredder residue that must be landfilled. *Bottom:* BMW has designed the Z1 Roadster so that external body panels and fascia can be easily removed from the automobile frame and subsequently recycled.

Box 1-A—Design and Materials Management in the Auto Industry

When an old car is junked, it is often first sent to a dismantler, who removes any parts that can be resold, as well as the battery, tires, gas tank, and operating fluids. The hulk is then crushed and sent to a shredder, which tears it into fist-sized chunks that are subsequently separated to recover the ferrous and nonferrous metals.

Presently, about 75 percent by weight of materials in old automobiles (including most of the metals) are recovered and recycled. The remaining 25 percent of the shredder output, consisting of one-third plastics (typically around 220 pounds of 20 different types), one-third rubber and other elastomers, and one-third glass, fibers, and fluids, is generally landfilled. In the United States, this shredder "fluff" amounts to about 1 percent of total municipal solid waste. Sometimes, the fluff is contaminated with heavy metals and oils, or other hazardous materials.

As automakers continue to search for ways to improve fuel efficiency and reduce manufacturing costs, the plastic content of cars is expected to increase. This will not only increase the amount of shredder fluff sent to landfills, it threatens the profitability of shredder facilities, which currently depend on metals recovery to make money.

In Germany, the landfilling of old automobile hulks and the shredder residues from automobile recycling operations is a growing problem. The German Government has proposed legislation that would require automakers to take back and recycle old automobiles at the end of their lifetime. This has stimulated German automakers to explore fundamental changes in automobile design that could result in more efficient materials management. These changes would involve new relationships among auto manufacturers, dismantlers, and materials suppliers.

To avoid dealing with the auto hulks themselves, the automakers propose to take better advantage of the existing infrastructure for auto recycling. Manufacturers will design cars that can be more cheaply disassembled, and will educate dismantlers on how to efficiently remove plastic parts. They will encourage their material suppliers to accept recovered materials from dismantlers, and will specify the use of recovered materials in new car parts, thus "closing the loop."

Green automobile design within this new framework of coordinated materials management has a very different character than auto design within an isolated firm. Instead of just thinking about how to design a fender or bumper using 10 percent less material, the designer also thinks about how the fender or bumper can be constructed from materials that can be co-recycled, and readily separated from the car body.

Several German companies, including BMW and Volkswagen, have begun to explore this system-oriented approach. BMW recently built a pilot plant in Bavaria to study disassembly and recycling of recovered materials, and Volkswagen AG has constructed a similar facility. The goal of the BMW facility is to learn to make an automobile out of 100 percent reusable/recyclable parts by the year 2000. In 1991, BMW introduced a two-seat roadster model with plastic body panels designed for disassembly and labeled as to resin type so they may be collected for recycling.

Interest in improving materials management in the auto industry is not limited to Europe. Japan's Nissan Motor Co. has announced research programs to explore design for disassembly, to reduce the number of different plastics used, to label those plastics to facilitate recycling, and to use more recovered materials in new cars. In the United States, Ford, Chrysler, and General Motors plan to label plastic components to identify the polymers, and have recently established a consortium with suppliers and recyclers (called the Vehicle Recycling Partnership) to address the recycling issue.

Autos are already one of the most highly recycled products in the United States. This success is largely due to the efficiency of shredder technology; a single facility can process up to 1,500 hulks per day. This level of productivity is not consistent with labor-intensive disassembly operations. Although research on recycling automotive plastics is ongoing, it is not yet economically feasible to separate and recycle these materials, even when avoided landfill tipping fees are included. Thus, it seems clear that a change in materials management in the U.S. auto industry is unlikely to emerge without substantial new economic or regulatory incentives.

SOURCE: Office of Technology Assessment, 1992.

Figure 1-5—Eco-labels Around the World

Canada (Environmental Choice)

Nordic Countries (White Swan)

West Germany (Blue Angel)

Japan (EcoMark)

United States (Scientific Certification Systems)*

United States (Green Seal)

Eco-labels are intended to identify environmentally preferred products for consumers. Above are government-sponsored labels from four foreign programs and two private U.S. labels.

*NOTE: The SCS label will provide comparative data on environmental attributes (see figure 4-1).

has no national program, but two private labeling efforts are underway (figure 1-5).

Product packaging, perhaps the most visible component of the post-consumer waste stream,

continues to be the target of control measures that include bans, taxes, deposits, and recycling requirements. One of the most dramatic initiatives is Germany's Packaging Waste Law, which gives manufacturers and distributors the responsibility for recovering and recycling their own packaging wastes. In fact, the idea of shifting the burden of dealing with discarded products from municipalities to manufacturers appears to be gaining momentum in Europe, and may soon be extended to durable goods, such as household appliances and automobiles. This statutory coupling of manufacturing with post-consumer recycling is forcing manufacturers—including U.S.-based manufacturers with subsidiaries in Europe—to change the way they design products.

The European Community is wrestling with the problem of harmonizing the different environmental product standards and recycling laws of member countries with the approach of the Single Market in 1992. These laws have proved contentious in the past, and harmonization is not yet in sight. Recent controversies over whether countries can restrict imports of goods deemed harmful to health or the environment, or whether such restrictions constitute nontariff barriers to trade, suggest that the harmonization of international environmental product policies is becoming a thorny problem that will have to be resolved in future negotiations under the General Agreement on Tariffs and Trade (GATT) and other international agreements.[10]

Many States within the United States are also enacting policies aimed at reducing the environmental impacts of products. These measures include mandating industry plans to reduce their use of toxic chemicals, mandating the disclosure of the use of hazardous chemicals in products, and establishing standard definitions for advertisers' use of terms like "recycled." States have also enacted some targeted product control measures such as recycled content requirements for newspaper, bans and taxes on specific packages, mandated manufacturer take-back of batteries, and tax incentives for recycling. The lack of uniform Federal environmental standards for products is alarming to industry, which fears having to satisfy 50 different State regulations. This prospect is especially of concern for products distributed through national networks.

---

[10] For an overview of the issues, see the U.S. Congress, Office of Technology Assessment, *Trade and Environment: Conflicts and Opportunities*, OTA-BP-ITE-94 (Washington, DC: U.S. Government Printing Office, May 1992).

The United States cannot be said to be "behind" other countries in the development of environmental policies that encourage green product design. Indeed, many European countries look enviously at U.S. environmental policies such as auto emissions standards, or the timetable for phaseout of CFC production and use, which are among the most aggressive in the world. Some U.S. chemical companies are acknowledged world leaders in waste prevention techniques.

It is more accurate to say that the U.S. approach differs substantially from the approaches being taken abroad, and these differences could create conflicts in the future. Whereas some of the "greener" European countries (especially Germany, the Netherlands, and the Nordic countries) increasingly focus on the environmental attributes of products at the national level, U.S. policies continue to focus on regulating industrial waste streams. Except in cases where products pose a clear threat to human health (e.g., some pesticides, PCBs, leaded gasoline and paint), the Federal Government has been reluctant to regulate the environmental attributes of products directly. For example, the Resource Conservation and Recovery Act regulates "hazardous" industrial waste closely, but delegates the primary responsibility for product disposal and "non-hazardous" solid waste management to the States.[11]

OTA finds no compelling reason for U.S. policies to necessarily imitate the product control policies in Germany or other countries (although monitoring the implementation of these initiatives could provide valuable lessons for the United States). In fact, many observers believe that some of the more extreme measures, such as Germany's mandatory take-back provisions for packaging waste, will prove to be costly and difficult to implement.

Nevertheless, the rapid evolution of environmental product policy, both in the States and abroad, suggests that the Federal Government needs to become more strongly involved for two reasons: 1) to keep abreast of technology and policy developments, and 2) to help shape policies that reduce barriers to interstate commerce and international trade. Options for greater Federal involvement are discussed below.

SHAPING ENVIRONMENTAL POLICIES THAT ENCOURAGE GREEN DESIGN

Some U.S. companies argue that existing market forces and environmental regulations are sufficient to encourage green design. They tend to view new environmental constraints on the design of products as a threat to their competitiveness and a drag on economic growth, especially during an economic recession. These companies are reluctant to redesign established products to achieve environmental benefits that have little value or visibility to their customers.

In fact, companies already have a number of incentives to move toward green design. By reducing the quantity of materials used in products, they can reduce manufacturing costs; by reducing the hazardous material content of products, they can reduce the rising costs of pollution control, waste disposal, and potential liability. There are also opportunities to gain consumer loyalty by enhancing the environmental attributes of their products. These incentives are already having an effect on the way many companies do business: for example, less toxic substitutes for heavy metals are being adopted in such products as inks, paints, and batteries; environmental advertising is being used to sell a range of products from gasoline to fabric softener; and more and more companies have recognized the linkages between improved product quality and improved environmental quality.

Even if Congress takes no further action, these incentives can be expected to continue in the future. For example, as permitted landfill capacity continues to shrink, waste disposal costs should increase, providing companies with greater incentives to reduce their wastes. Implementation of tougher emissions standards under the Clean Air Act Amendments of 1990 will increase pressures on companies to reduce their use of hazardous solvents and other volatile organic compounds. Various States will no doubt continue to pass legislation to regulate the environmental attributes of products and waste streams. And as consumers become more attuned to environmental concerns, they will increasingly de-

---

[11] However, the RCRA reauthorization debate has involved a number of new proposals that would establish national requirements for product design, including mandatory recycled content, reduced toxic chemical content and government procurement of recycled products.

Photo credit: "Selling Green." Copyright 1991 by Consumers Union of U.S., Inc., Yonkers, NY 10703-1057. Reprinted by permission from CONSUMER REPORTS, October 1991.

This collage of fictional packages illustrates the trend toward environmental marketing.

mand that manufacturers take more responsibility for the environmental impacts of their products.

But OTA finds that there are four specific areas of need that existing market forces or regulations do not adequately address, and that are uniquely the responsibility of the Federal Government:

- *Research*—At present, policymakers don't know what materials or waste streams are of greatest concern, or about how product designs might be changed to address them most effectively. Private companies have no incentive to conduct this research.

- *Credible information for consumers*—Surveys show that consumers are interested in green products, but most don't know what is "green." As discussed above, defining what's green is a multidimensional problem. In the absence of Federal action to establish consistent ground rules defining terms and measurement methods, the growing interest of consumers could become dissipated in confusion and skepticism.

- *Market distortions and environmental externalities*—Despite the existing incentives for green design noted above, critics of present consumption patterns argue that important market distortions and environmental externalities remain that encourage inefficient use of materials and energy. Failure to internalize

these environmental costs into design and production decisions can make environmentally sound choices seem economically unattractive.

- *Coordination and harmonization*—OTA found that several research projects related to green design are being sponsored by various Federal agencies and offices, but that there is little or no coordination among them. And unlike its major competitors, the United States has no institutional focus at a national level for addressing environmental product policy.

Current Federal Efforts Related to Green Design

OTA identified a number of ongoing Federal activities that partially address these needs (table 1-1). EPA is most directly involved. For example, its Office of Research and Development has several projects underway to develop generic guidelines for green design. However, there are relevant projects scattered through several other agencies, including the Department of Energy (DOE) and the National Science Foundation.

Several recent initiatives could help to remove some of the barriers to green design that exist in current Federal rules and regulations. In October 1991, President George Bush signed Executive Order 12780, the Federal Recycling and Procurement Policy, which requires Federal agencies to increase recycling and waste reduction efforts, and to encourage markets for recovered materials by favoring the purchase of products with recycled content. The Department of Defense has issued recent directives emphasizing waste prevention through the acquisition process and through military specifications and standards. Some 40,000 military specifications requiring the use of hazardous materials are currently under review. These initiatives will help to create markets for green products.

There are also several ongoing activities that could improve the quality of information available to consumers and citizen groups. EPA and the Federal Trade Commission (FTC) are developing guidelines for advertisers' use of environmental terms such as "recycled." National standards for use of these terms in advertising can give consumers confidence that a product advertised as "source-reduced" or "recycled" is really better for the environment. In the Pollution Prevention Act of

Table 1-1—Federally Funded Programs Related to Green Design

Agency/office	Program/activity	Comments
Department of Energy		
Office of Industrial Technologies	Industrial Waste Reduction Program	This research and development program aims to identify priority industrial waste streams, assess opportunities for addressing these waste streams through redesigning products and production processes, and technology transfer from national laboratories.
Environmental Protection Agency		
Office of Research and Development	Environmental Resource Guide	Contracted to the American Institute of Architects, this project will provide information to architects on the life-cycle environmental impacts of construction materials.
	Dynamic Case Studies on Environmentally Advanced Product Design	Contracted to the Resource Policy Institute in Los Angeles and the Product Life Institute in Geneva, this project will explore case studies involving green product design.
	Life Cycle Assessment Methodology	Contracted to Battelle, this project will develop standard methodologies for conducting product life-cycle assessments.
	Clean Products Case Studies	Contracted to INFORM Inc., this project will provide case studies of green design, especially the reduced use of toxic substances in products.
	Safe Substitutes	Contracted to the University of Tennessee, this project will identify priority toxic chemicals and evaluate possible substitutes.
	Life Cycle Design Guidance Manual: Environmental Requirements and the Product System	Contracted to the University of Michigan, this manual will explore how designers can incorporate life-cycle information into their designs.
	National Pollution Prevention Center	Located at the University of Michigan, this center is developing waste prevention information modules for industrial and engineering design courses.
	American Institute for Pollution Prevention	In association with the University of Cincinnati, the institute serves as a liaison to a broad cross-section of industry, with projects involving four aspects of waste prevention: education, economics, implementation, and technology.
Office of Pollution Prevention and Toxics	Design for the Environment	Proposed program to gather, coordinate, and disseminate information on green design.
National Science Foundation	Engineering Design Research Center	Located at Carnegie Mellon University, the center is organizing a program to explore methods for green design.

SOURCE: Office of Technology Assessment.

1990, Congress required manufacturers who report their releases of toxic chemicals for the Toxic Release Inventory (TRI)[12] to also report how these releases were affected by product and process redesign. When this provision is implemented, it could become a valuable source of information in an area where little information currently exists: how product design choices affect industrial waste streams.

In the short term, Congress can make a good start toward encouraging green design by using its oversight powers to ensure that these ongoing activities are carried through to their conclusion, and that the provisions of the Pollution Prevention Act are implemented expeditiously.

Long-Term Options

In the longer term, Congress may wish to address the needs identified above more directly. These needs are discussed in greater detail below.

Research

Of critical importance is to identify what materials and products pose the greatest risks to human health and the environment. Without this information, Congress cannot intelligently set priorities for

---

[12] As required under Title III of the Superfund Amendments and Reauthorization Act of 1986.

environmental policy. Congress could direct EPA and DOE to identify a short list of priority materials, products, and waste streams; identify areas where additional data are needed to assess their health and environmental impacts; and develop quantitative models showing how these high-risk materials flow through the economy.

Research is also needed to develop techniques for measuring the environmental impact of products and systems, to better understand how the business climate and corporate culture affect product design decisions vis-a-vis the environment, and to explore the costs and benefits of various policy options such as manufacturer take-back requirements.

Credible Information for Consumers

As discussed above, national standard definitions for advertisers' use of environmental terms could alleviate consumer confusion associated with current environmental claims. An important goal for the future will be to determine how to credit products that feature waste prevention in regulations and government procurement programs that are currently focused on recycling and recycled content. For example, should waste prevention be measured with respect to waste generated in some previous base year, or with respect to other comparable products in the current year?

The United States has two small, private eco-labeling efforts underway. One potential concern is that a variety of private labels based on different appraisal methods could lead to confusion about which products are actually better for the environment. To address this problem, Congress could designate EPA to develop standards for the certification of private eco-labeling programs. This might give consumers confidence that products carrying certified eco-labels are in fact better for the environment. Alternatively, it could appoint a blue-ribbon commission to oversee the establishment of an independent, national eco-labeling program similar to those of other countries.

Where the disclosure of public information on industrial waste streams has been mandated by Congress, e.g., through the Toxic Release Inventory, this has proven to be a powerful motivation for companies to change their designs and manufacturing processes. Expanded industry reporting requirements under TRI could improve the information on materials flows available to public interest groups

Credit: Wayne Stayskal, Tampa Tribune

and ultimately to consumers. This might involve expanding the number of reportable chemicals, the types of industries required to report, or expanding reporting requirements themselves to include the actual *use* of priority chemicals in products and processes, not just the *release* of these chemicals to the environment. However, unless these requirements are narrowly targeted on chemicals or materials of special concern (see research needs above), they would significantly increase industry's reporting costs, and might not result in a commensurate reduction of environmental risk.

Market Distortions and Environmental Externalities

Products have environmental impacts at every stage of their life cycle. Yet, many of these are not accounted for in the prices of materials and products. On the production side, there are government subsidies or special tax treatment for the extraction of virgin materials (e.g., below-cost timber sales and mineral depletion allowances), and many "nonhazardous" industrial solid wastes (e.g., mine tailings or manufacturing wastes that are managed on-site) with significant environmental impacts are not regulated at the Federal level. On the consumption side, consumers often do not pay the full environmental costs of products that are consumed or dissipated during use (e.g., gasoline, cleaners, agricultural chemicals), or the full cost of solid waste disposal.

Table 1-2 presents a menu of regulatory and market-based incentives that have been proposed to address environmental problems associated with the flow of goods and materials through the economy.

Table 1-2—Policy Options That Could Affect Materials Flows

Life-cycle stage	Regulatory instruments	Economic instruments
Raw material extraction and processing	Regulate mining, oil, and gas non-hazardous solid wastes under the Resource Conservation and Recovery Act (RCRA). Establish depletion quotas on extraction and import of virgin materials.	Eliminate special tax treatment for extraction of virgin materials, and subsidies for agriculture. Tax the production of virgin materials.
Manufacturing	Tighten regulations under Clean Air Act, Clean Water Act, and RCRA. Regulate non-hazardous industrial waste under RCRA. Mandate disclosure of toxic materials use. Raise Corporate Average Fuel Economy Standards for automobiles. Mandate recycled content in products. Mandate manufacturer take-back and recycling of products. Regulate product composition, e.g., volatile organic compounds or heavy metals. Establish requirements for product reuse, recyclability, or biodegradability. Ban or phase out hazardous chemicals. Mandate toxic use reduction.	Tax industrial emissions, effluents, and hazardous wastes. Establish tradable emissions permits. Tax the carbon content of fuels. Establish tradable recycling credits. Tax the use of virgin toxic materials. Create tax credits for use of recycled materials. Establish a grant fund for clean technology research.
Purchase, use, and disposal	Mandate consumer separation of materials for recycling.	Establish weight/volume-based waste disposal fees. Tax hazardous or hard-to-dispose products. Establish a deposit-refund system for packaging or hazardous products. Establish a fee/rebate system based on a product's energy efficiency. Tax gasoline.
Waste management	Tighten regulation of waste management facilities under RCRA. Ban disposal of hazardous products in landfills and incinerators. Mandate recycling diversion rates for various materials. Exempt recyclers of hazardous wastes from RCRA Subtitle C. Establish a moratorium on construction of new landfills and incinerators.	Tax emissions or effluents from waste management facilities. Establish surcharges on wastes delivered to landfills or incinerators.

SOURCE: Office of Technology Assessment.

These options are organized according to their point of greatest impact in the product life cycle. Each could have an impact on product design, but an analysis of the design implications of these options is beyond the scope of this report. However, OTA offers three guiding principles that policymakers should consider as they evaluate these options (see below).

Coordination and Harmonization

Green design is a multidisciplinary subject that does not fit comfortably within the mission of any single Federal agency. For instance, EPA is organized around regulatory responsibilities for protecting air, water, and land; its technical expertise in design and manufacturing areas is slight. The Department of Commerce (DOC), on the other hand, is concerned with the competitiveness of industrial sectors, but has little environmental expertise. DOE's

national labs have considerable experience that could be brought to bear on improving the efficiency of industrial energy use and waste prevention, but environmental quality has not traditionally been a part of DOE's mission. The story is much the same with other agencies.

As described here, **green design refers not to a rigid set of product attributes, but rather to a decision process whose objectives depend upon the specific environmental problems to be addressed.** This suggests that the most meaningful way in which the Federal Government can encourage green design is through multiagency initiatives organized around particular environmental problems, policy issues, or industrial sectors. These collaborations are beginning to be formed on an ad hoc basis. For instance, EPA is working with the Department of Agriculture to promote waste prevention in agricultural chemical use. EPA, DOE, and DOC are collaborating in a joint grant program with States to fund research on reducing the environmental impacts of industrial processes.[13] These efforts are a start; however, much more could be done in the following areas:

- *Promoting information exchange*—Current mechanisms for disseminating information on relevant research activities in various agencies are only partially effective. A central repository containing information on all relevant Federal research activities would be helpful.
- *Promoting system-oriented design solutions*— A greener transportation sector, for example, may involve not only improved vehicle fuel efficiency, but better management of materials used in automotive, rail, aviation, etc., as well as changes in urban design. A multiagency perspective could provide a more holistic analysis of total sectoral issues, through forums, grant programs, etc.
- *Harmonizing State and Federal environmental product policies*—Policy guidance is needed to help define the circumstances under which Federal standards preempting State and local product control laws may be justified, and where they are not.
- *Coordinating policy development on international aspects of the environment, U.S. competitiveness, trade, and technology*—At present, responsibility for development of U.S. policy in these areas is not clearly defined, with each Federal agency having its own agenda.

To address these needs, Congress could:

- Provide funding for a central electronic network listing current Federal research projects, case studies, and bibliographic materials relating to green design.
- Use its oversight powers to clarify which agencies have lead responsibility for policy development on interstate and international aspects of U.S. environmental product policy.
- Ensure that green design considerations are integrated into the charter of any new environmental or technological organizations now being proposed before Congress, such as the National Institutes for the Environment or a Civilian Technology Agency.

In the end, the institutional details are less important than a recognition on the part of Congress and the Administration that Federal leadership is needed to take advantage of opportunities such as green design that do not fall neatly within the mission of any single agency.

GUIDING PRINCIPLES FOR POLICY DEVELOPMENT

The discussion above suggests three general principles that Congress can use to develop environmental policies that encourage, rather than inhibit, green design.

Principle 1: Identify the root problem and define it clearly.

One of the biggest challenges in selecting a policy strategy is clearly defining the problem to be addressed. One difficulty is that products and waste streams have multiple environmental impacts that cannot be easily disentangled. Policymakers may be concerned with the quantity of a particular waste stream, its toxicity, or persistence in the environment. Policies aimed at solving one problem may have unintended negative effects on another; for example, requiring automobiles to be made from currently recyclable materials could adversely affect their fuel efficiency. Inevitably, tradeoffs and value judgments must be made as to which environmental impacts are the most important.

Disagreements about how to define the environmental problem may also reflect more fundamental philosophical differences. Industry tends to frame the problem in terms of reducing the quantity of waste destined for disposal, while environmental groups often focus on threats to natural resources and ecological "sustainability." These different problem statements lead to different policy prescriptions and different ideas about what constitutes green design. Clearly defining the problem to be addressed can help to elevate the level of debate and to identify possible areas of compromise.

In the absence of a clearly defined problem, it becomes easy to confuse means and ends. Proxies for environmental quality, such as recycling, can come to be perceived as ends in themselves, rather than as one of several strategies for reducing solid

---

[13] The program is called National Industrial Competitiveness through Efficiency: Energy, Environment, Economics (NICE[3]).

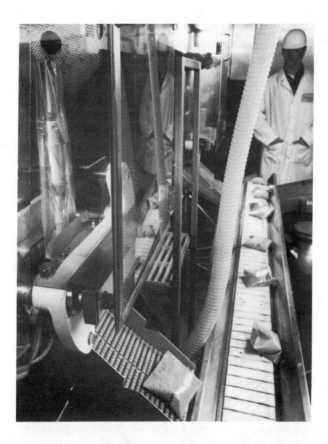

Photo credit: Dupont Magazine (July/August 1991)

A thin plastic milk pouch uses less material than a traditional paperboard carton, and takes up less space in a landfill.

waste. By mandating that products and packaging contain a minimum recycled content, for instance, Congress would certainly encourage product designers to use recovered materials in packaging; but this would not necessarily result in less waste overall. Perversely, this could even lead to more waste, especially if designs featuring waste prevention are thereby discouraged. If the objective is to reduce the amount of solid waste generated, MSW policies and government procurement programs should make allowances for product designs that feature waste prevention.

Defining the problem properly must entail some consideration of environmental risks. OTA finds that policymakers currently lack critical information on how materials flow through the economy and about the relative risks of different materials and products. For example, 10 States have passed legislation banning the use of heavy metals in packaging, even though this source contributes only 4 to 7 percent of heavy metals in landfills and incinerators. Without research to develop information on materials flows and relative risks, resources are likely to be directed toward the most visible problems, rather than those that pose the greatest environmental risks.

Principle 2: Give designers the maximum flexibility that is consistent with solving the problem.

Materials technology options are proliferating rapidly, and product impacts on the environment are nearly always multidimensional. Policies should be crafted to give designers as much flexibility as possible to find the best solutions, within a framework that protects human health and the environment. Rigid Federal mandates that impose predetermined design solutions (such as bans on the use of certain materials) are likely to be inefficient, and should be avoided if possible. Flexibility can be achieved through a variety of means, including flexible regulations, economic instruments, and negotiated voluntary agreements with industry.

One tradeoff for increasing policy flexibility is likely to be increased costs of policy monitoring and enforcement. For example, verifying compliance with a ban on the use of a given material or chemical requires less information than verifying compliance with voluntary agreements or flexible regulations. For this more flexible approach to work, the cost of demonstrating compliance will probably have to be borne by industry.

Principle 3: Encourage a systems approach to green design.

If policymakers focus exclusively on addressing the environmental attributes of products, as opposed to the systems in which products are manufactured, used, and disposed, they are likely to miss the biggest opportunities for green design.

A system-oriented design approach can be encouraged either directly by regulation, or indirectly through economic incentives. Recycled content regulations or manufacturer take-back requirements are examples of a regulatory coupling between manufacturing and waste management. The proposal of the German Government to require auto manufacturers to take back and recycle their cars, for example, has stimulated the German automakers to rethink the entire ''ecology'' of auto production and disposal (box 1-A). This approach may be more appropriate for high-value, durable products than for nondurable or disposable products.

An alternative to take-back regulations is to indirectly encourage designers to take a systems approach by using economic instruments to internalize the costs of environmental services. This approach would rely on market forces to sort out what new interfirm relationships make sense economically, while giving designers the flexibility to design products with the best combination of cost, performance, and least environmental impact. For example, a substantial carbon tax on fuels could have a dramatic impact on the systems by which products are manufactured, distributed, and disposed, because fuels are consumed at every stage of the product life cycle.

CONCLUSION

Green product design offers a new way of addressing environmental problems. By recasting pollution concerns as product design challenges, and particularly by encouraging designers to think more broadly about production and consumption systems, policymakers can address environmental problems in ways that would not have been apparent from a narrow focus on waste streams alone.

The flow of materials and products through the world economy has a critical influence on both economic growth and the environment. These flows are determined in part by design decisions. Therefore, policymakers should strive to make green product design an integral part of strategies to improve competitiveness and environmental quality. OTA's analysis suggests that simply providing information to designers and consumers about the environmental impacts of products and waste streams is not enough. To move ahead, the environmental costs of production, consumption, and disposal should be accounted for at each stage of the product life cycle. The challenge to policymakers is to choose a mix of regulatory and economic instruments that target the right problems and give designers the flexibility to find innovative, environmentally elegant solutions.

Chapter 2

Environmental Aspects of Materials Use

Contents

Figures

Environmental Aspects of Materials Use

The world economy is consuming resources and generating wastes at unprecedented rates. In the past 100 years, the world's industrial production increased more than 50-fold,[1] releasing some materials to the environment at rates that far exceed releases occurring naturally. Human activities are estimated to release several times as much chromium, nickel, arsenic, and selenium to the atmosphere as do natural processes, and over 300 times as much lead.[2] Carbon dioxide levels in the atmosphere are increasing at a rate 30 to 100 times faster than the rate of natural fluctuations observed in the climatic record.[3]

The U.S. economy is among the most material intensive economies in the world, extracting more than 10 tons (20,000 lb) of "active" material per person from U.S. territories each year.[4] Most of this material becomes waste relatively quickly. By one estimate, only 6 percent of this active material is embodied in durable goods; the other 94 percent is converted into waste within a few months of being extracted.[5]

These statistics on material flows do not directly measure the increased risks to human health or ecosystems, but recent experience with ozone depletion and the threat of global warming indicates that such explosive growth in materials flows could have profound and possibly irreversible environmental consequences. This growth is expected to continue; by the middle of the next century, the world population is expected to double,[6] and the global economy could be five times as large.[7]

The environmental risks posed by increasing materials flows can be addressed both by improving industrial efficiency and by substituting less damaging materials for those currently in use. For example, substitutes for chlorofluorocarbons (CFCs) are becoming available as the production of these ozone-depleting chemicals is phased out. However, such actions tend to be taken only in direct response to government regulations or after some specific environmental problem has reached threatening dimensions. Industrial production decisions have generally not considered the environmental impacts of materials and process choices in a proactive way.

HOW PRODUCT DESIGN AFFECTS THE ENVIRONMENT

Product design decisions have impacts on the environment at each stage of the product life cycle, from extraction of raw materials to final disposal (figure 2-1). Ideally, one would like design decisions to take account of both the "downstream" impacts (product use and disposal) as well as the "upstream" impacts (materials extraction, processing, and manufacturing).

The most publicly visible environmental impacts associated with products are the "downstream" impacts, particularly municipal solid waste (MSW). U.S. households and commercial establishments generate about 4 pounds of trash per person each day. In 1988, the United States generated some 180 million tons of MSW (figure 2-2). This amount is expected to increase by about 1.5 percent per year, resulting in total MSW generation of over 215

[1] Based on data from W.W. Rostow, *The World Economy: History and Prospects* (Austin, TX: University of Texas Press, 1978), pp. 48-49.

[2] James N. Galloway, J. David Thornton, Stephen A. Norton, Herbert L. Volchok, and Ronald A.N. McLean, *Atmospheric Environment* 16(7):1678, 1982. See also Robert U. Ayres, "Toxic Heavy Metals: Materials Cycle Optimization," *Proceedings of the National Academy of Sciences*, vol. 89, No. 3, Feb. 1, 1992, pp. 815-820.

[3] U.S. Congress, Office of Technology Assessment, *Changing by Degrees: Steps To Reduce Greenhouse Gases*, OTA-O-482 (Washington, DC: U.S. Government Printing Office, February 1991), p. 45.

[4] "Active" material includes food, fuel, forestry products, ores, and nonmetallics. It excludes inert construction materials such as sand, gravel, and stone, as well as atmospheric oxygen and fresh water. Robert U. Ayres, "Industrial Metabolism," *Technology and Environment* (Washington, DC: National Academy Press, 1989), p. 25.

[5] Ibid, p. 26.

[6] United Nations, Department of International Economic and Social Affairs, *Long-Range World Population Projections: Two Centuries of Population Growth 1950-2150* (New York, NY: United Nations, 1992), p. 14.

[7] George Heaton, Robert Repetto, and Rodney Sobin, *Transforming Technology: An Agenda for Environmentally Sustainable Growth in the 21st Century* (Washington, DC: World Resources Institute, April 1991), p. 1.

Figure 2-1—Stages of the Product Life Cycle

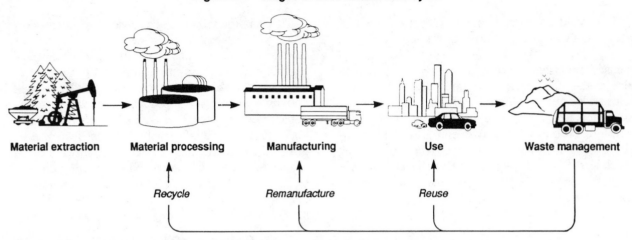

| Material extraction | Material processing | Manufacturing | Use | Waste management |

Recycle   Remanufacture   Reuse

Environmental impacts occur at all stages of a product's life cycle. Design can be employed to reduce these impacts by changing the amount and type of materials used in the product, by creating more efficient manufacturing operations, by reducing the energy and materials consumed during use, and by improving recovery of energy and materials during waste management.

SOURCE: Adapted from D. Navin chandra, The Robotics Institute, Carnegie Mellon University, personal communication, March 1992.

million tons by the year 2000.[8] Landfills in many States are reaching their permitted capacity, and there is increasing public opposition to siting new waste management facilities. One major reason for this opposition is concern about toxic materials released from these facilities, e.g., when batteries are incinerated, or when household hazardous waste is placed in landfills.

Less visible but potentially more serious environmental impacts occur during raw material extraction, material processing, and product manufacturing. U.S. industry generates some 700 million tons of "hazardous waste" and some 11 *billion* tons of "non-hazardous" solid waste.[9] Although the weight of industrial and municipal solid waste cannot be compared directly,[10] industrial wastes dwarf municipal solid wastes in their quantity and environmental impact (see figure 2-3).[11] Product design decisions have a direct influence on the manufacturing component of these wastes (about 6.5 billion tons).

Finally, some of the most serious environmental releases occur during the actual use of the product. This is particularly true of products that are consumed or dissipated during their use (e.g., volatile solvents and propellants, fuels, cleaners and paints, and agricultural fertilizers and pesticides).[12] Prime examples are CFCs used as coolants, solvents, and

---

[8] U.S. Environmental Protection Agency, *Characterization of Municipal Solid Waste in the United States: 1990 Update*, June 1990, pp. ES-3 and 75; U.S. Congress, Office of Technology Assessment, *Facing America's Trash: What Next for Municipal Solid Waste*, OTA-O-424 (Washington, DC: U.S. Government Printing Office, October 1989).

[9] The terms *hazardous* and *non-hazardous* are defined by the Resource Conservation and Recovery Act (RCRA), Subtitles C and D, respectively. Industrial solid wastes not defined as hazardous under Subtitle C fall under Subtitle D of RCRA. See U.S. Congress, Office of Technology Assessment, *Managing Industrial Solid Wastes From Manufacturing, Mining, Oil and Gas Production, and Utility Coal Combustion—Background Paper*, OTA-BP-O-82 (Washington, DC: U.S. Government Printing Office, February 1992), pp. 4-15.

[10] Up to 70 percent of the weight of industrial solid waste (which includes mining, oil and gas, and manufacturing wastes) consists of wastewater contained in sludges and aqueous solutions.

[11] As figure 2-3 indicates, industrial wastes clearly dwarf MSW by weight. In terms of environmental impact, even if all MSW were "hazardous," industrial "hazardous" wastes alone would still be three times as large. Furthermore, some of the "non-hazardous" wastes do not differ substantially from wastes designated as "hazardous" under the Resource Conservation and Recovery Act, or they may exhibit other characteristics of concern (see OTA, *Managing Industrial Solid Wastes From Manufacturing, Mining, Oil and Gas Production, and Utility Coal Combustion—Background Paper*, op. cit., footnote 9, p. 12).

[12] For example, the energy required to manufacture motor vehicle components and assemble those components into finished vehicles totaled about 1.5 quadrillion Btu (quads) in 1985. However, the fuel *used* in motor vehicles totaled more than 10 times that amount. Sources: U.S. Congress, Office of Technology Assessment, *Energy Use and the U.S. Economy*, OTA-BP-E-57 (Washington, DC: U.S. Government Printing Office, June 1990), p. 3; U.S. Department of Energy, Energy Information Administration, *Annual Energy Review 1990*, DOE/EIA-0384(90) (Washington, DC: U.S. Government Printing Office, May 1991), p. 53.

Figure 2-2—Municipal Solid Waste Management (1960-90)

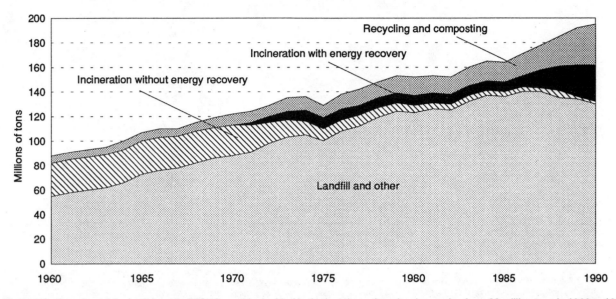

The generation of municipal solid waste (MSW) roughly doubled in the last three decades, increasing from 88 million tons in 1960 to 196 million tons in 1990 (population growth accounts for roughly one-third of the increase). Techniques of managing MSW changed somewhat during this period: recycling and composting increased substantially; total incineration remained roughly constant, but nearly all incineration now occurs with energy recovery; landfilling increased for most of the period, but has leveled off in recent years.

SOURCE: Franklin Associates, Ltd., personal communication, August 1992.

blowing agents, which have been linked to depletion of the stratospheric ozone layer.

In some cases, the environmental releases from products can be larger than those from the associated industrial processes. For example, heavy metals (e.g., mercury, lead, cadmium, chromium, and nickel) are among the most toxic constituents of industrial wastes.[13] Although complete data on industrial inputs and outputs of heavy metals are scarce, data collected under New Jersey's Worker and Community Right to Know Act of 1983 indicate that most heavy metals that enter industrial processes end up in products, not industrial wastes. In 1990, for example, at least 55 to 99 percent of industrial inputs of these five heavy metals were used in products, depending on the metal.[14] While

some of these products are recovered and recycled,[15] much of the heavy metal content of products is released into the environment (e.g., in paints and coatings) or enters landfills and incinerators (e.g., in plastics).

TRENDS IN MATERIALS USE

During this century, dramatic changes have occurred in the nature of the materials Americans use to manufacture products. Figure 2-4 shows the consumption of different classes of materials in the United States between 1900 and 1989.[16] The top half of the figure shows consumption in absolute terms. During the past 90 years, consumption of raw materials derived from agricultural and forestry commodities has grown slowly. In contrast, there has been dramatic growth in consumption of raw

---

[13] These five heavy metals are all targeted in the Environmental Protection Agency's 33/50 Program—an effort aimed at encouraging industry to voluntarily reduce releases of 17 priority chemicals 33 percent by the end of 1992 and 50 percent by the end of 1995. See U.S. Environmental Protection Agency, Office of Pollution Prevention and Toxics, *Pollution Prevention Resources and Training Opportunities in 1992*, January 1992, pp. 84-85.

[14] Data from Andrew Opperman, New Jersey Department of Environmental Protection and Energy, Community Right to Know Program, Bureau of Hazardous Substances Information, personal communication, August 1992.

[15] Recycling rates vary substantially by material. For example, more than 50 percent of lead is recycled, but nearly all cadmium is released into the environment.

[16] Figures 2-4a and 2-4b measure material consumption by *value* to allow aggregation of diverse material types such as bales of cotton, barrels of oil, tons of ore, and cubic feet of gas. The figure only includes materials consumed for uses other than food and fuel. Source: David Berry, Program Manager, Material Use Trends and Patterns, Bureau of Mines, U.S. Department of the Interior, personal communication, July 1992.

Figure 2-3—"Solid" Wastes as Defined Under the Resource Conservation and Recovery Act (RCRA)

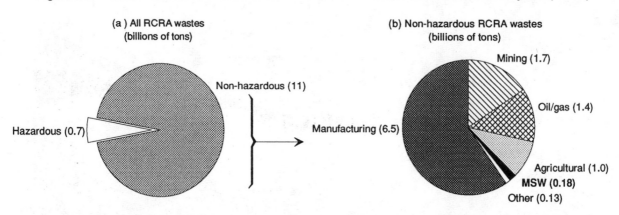

Much of the solid waste produced in the United States is not directly generated by consumers. Municipal solid waste, the focus of much public concern, represents less than 2 percent of all solid waste regulated under RCRA. In contrast, industrial activities produce about 700 million tons of hazardous waste (a) and about 11 *billion* tons of non-hazardous wastes (b).

NOTE: All numbers are estimates. The non-hazardous waste total has been rounded to reflect uncertainty. Much of the "solid" waste defined under RCRA, perhaps as much as 70 percent, consists of wastewater. The terms *hazardous* and *non-hazardous* refer to statutory definitions of Subtitles C and D of RCRA, respectively. The mining wastes shown in (b) exclude mineral processing wastes; the oil/gas wastes in (b) exclude produced waters used for enhanced oil recovery; the "other" category in (b) includes wastes from utility coal combustion.

SOURCE: Adapted from U.S. Congress, Office of Technology Assessment, *Managing Industrial Solid Wastes From Manufacturing, Mining, Oil and Gas Production, and Utility Coal Combustion*, OTA-BP-O-82 (Washington, DC: U.S. Government Printing Office, February 1992).

materials derived from ores and minerals (used in the production of steel, aluminum, and asbestos) and of raw materials derived from organic feedstocks (used in the production of plastics, fibers, petrochemicals, and asphalt).

The bottom half of figure 2-4 shows these shifts in comparative terms. In 1900, the majority of the raw materials consumed were derived from agricultural and forestry products. By the late 1980s, materials derived from ores and minerals constituted about 50 percent of all raw materials, up from only about 30 percent in 1900; materials derived from organic feedstocks (such as plastics, fibers, petrochemicals, and asphalt) comprised about 15 percent of the total, while in 1900 these materials practically did not exist.

A closer examination of these changes reveals three important trends in materials use: increasing variety, increasing efficiency, and increasing complexity. These trends are closely related, but each is significant in its own right.

Increasing Variety

Materials use has changed not only in terms of the relative amounts of different materials, but also in the variety of materials available. A century ago, U.S. industry utilized only about 20 elements of the periodic table; today, virtually all 92 naturally occurring elements are used.[17] Moreover, with advances in the understanding of the structure of physical matter, researchers have created thousands of chemical compounds and a broad array of novel materials. In chemicals alone, it is estimated that over 60,000 have been synthesized and roughly 10,000 are produced in commercial quantities.[18] About 1,000 new chemicals are introduced each year, and are incorporated into products as diverse as pharmaceuticals, superadhesives, and agricultural pesticides.[19]

Remarkable advances in structural materials technologies have led to the development of ceramics and composites that offer superior properties (e.g., high-temperature strength, high stiffness, and light weight) compared with traditional materials such as

[17] *Materials and Man's Needs: The History, Scope, and Nature of Materials Science and Engineering*, vol. I. (Washington, DC: National Academy of Sciences, September 1975), ch. 1, p. 2.

[18] Michael Shapiro, "Toxic Substances Policy," *Public Policies for Environmental Protection*, Paul R. Portney (ed.) (Washington, DC: Resources for the Future, 1990), p. 195.

[19] Ibid.

Figure 2-4—U.S. Raw Material Consumption (1900-89)

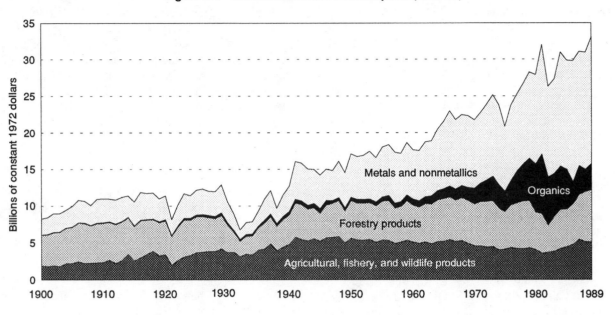

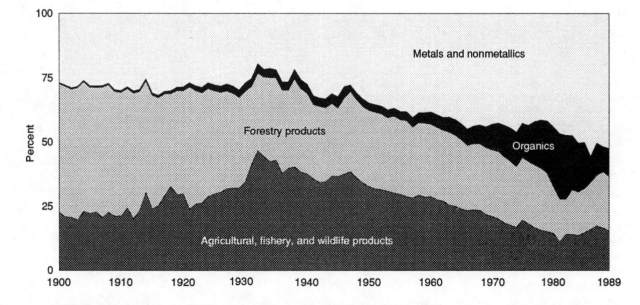

U.S. raw material consumption has changed dramatically in this century. In absolute terms (top), raw material consumption has increased by a factor of 4 (population has increased by roughly a factor of 3 during the same period). The largest increases were in materials derived from mining operations (metals and nonmetallic ores) and from organics (plastics and petrochemicals). In relative terms (bottom), the sources of raw materials consumed in the United States have gone from predominantly agriculture and forestry to predominantly mining and organics.

NOTE: The data measure material consumption by value to allow for aggregation of diverse material types. The data only include materials consumed for uses other than food and fuel.

SOURCE: Data prior to 1978 from Vivian E. Spencer, *Raw Materials in the United States Economy, 1900-1977* (Washington, DC: U.S. Department of Commerce, Bureau of the Census, 1980); data after 1978 from Bureau of Mines, op. cit., footnote 16.

steel and aluminum.[20] These materials can be engineered to have the precise properties required for a given application. Use of such designed materials can lead to higher fuel efficiency, lower assembly costs, and longer service life for many manufactured products.

Recent developments in materials technology are impressive in terms of both breadth and ingenuity. High-temperature superconductors offer the promise of extremely efficient electronic devices and power transmission systems.[21] Conductive plastics may someday combine the electrical characteristics of copper with the strength of steel, and may lead to lightweight batteries and electric motors.[22] "Smart" materials—materials that sense and react to changes in their operating environment—may result in helicopter rotors that stiffen in response to turbulence or temperature fluctuations, or shock absorbers that automatically adjust to changing driving conditions. Molecular beam epitaxy techniques allow semiconductor devices to be built atom by atom, suggesting the possibility of hand-held supercomputers and ultra-small, low-power lasers for use in communications.[23]

Even the profound impacts of these so-called "engineered materials" may eventually be overshadowed by biologically derived substances. By harnessing the enzymes of nature, an entirely new range of nontoxic, renewable, biodegradable, and biocompatible materials may be on the horizon.

Researchers are looking at how biopolymers might be used for applications as diverse as artificial skin, superabsorbants, dispersants, and as permeable coatings for agricultural seed. Several biologically derived polymers are already in production.[24] Ultimately, plants may be genetically programmed to produce plastic instead of starch.[25] Such developments could potentially reduce society's dependence on petroleum-based materials.

Increasing Efficiency

New processing technologies, more sophisticated materials, and improved product design have resulted in the more efficient use of materials. For example, an office building that can be built with 35,000 tons of steel today required 100,000 tons 30 years ago.[26] Similarly, aluminum cans today weigh 30 percent less than they did 20 years ago.[27] [28]

Traditional materials have been displaced in many applications by new light-weight materials such as high strength alloys and plastics. For example, telecommunications cables in the 1950s consisted mostly of steel, lead, and copper, with a small percentage of aluminum and plastics (figure 2-5). By the 1980s, the plastics content of cables had increased to more than 35 percent and the lead content had dropped to less than 1 percent. If polyethylene had not replaced lead as cable sheathing, AT&T's lead requirements might have approached a billion pounds annually.[29] The process of substitution continues: today, 2,000 pounds of

---

[20] See U.S. Congress, Office of Technology Assessment, *Advanced Materials by Design*, OTA-E-351 (Washington, DC: U.S. Government Printing Office, June 1988).

[21] U.S. Congress, Office of Technology Assessment, *High-Temperature Superconductivity in Perspective*, OTA-E-440 (Washington, DC: U.S. Government Printing Office, April 1990).

[22] "The Promise of Conductive Plastics," *EPRI Journal*, July/August 1991, pp. 5-13.

[23] U.S. Congress, Office of Technology Assessment, *Miniaturization Technologies*, OTA-TCT-514 (Washington, DC: U.S. Government Printing Office, November 1991).

[24] The chemical firm ICI is producing about 50 tons of polyhydroxybutyrate-valerate (PHBV) annually. See William D. Luzier, "Materials Derived From Biomass/Biodegradable Materials," *Proceedings of the National Academy of Sciences*, vol. 89, No. 3, Feb. 1, 1992, pp. 839-842.

[25] "In Search of the Plastic Potato," *Science*, Sept. 15, 1989, pp. 1187-1189.

[26] Milton Deaner, President, American Iron & Steel Institute, presentation at the Bureau of Mines Forum on Materials Use, Washington, DC, Sept. 17, 1991.

[27] S. Donald Pitts, Vice President, Aluminum Association, presentation at the Bureau of Mines Forum on Materials Use, Washington, DC, Sept. 17, 1991.

[28] This increasing materials efficiency is a component of what some observers call "dematerialization"—a decrease in the materials consumed per unit of GNP. See Robert H. Williams, Eric D. Larson, and Marc H. Ross, "Materials, Affluence, and Industrial Energy Use," *Annual Review of Energy* 12:99-144, 1987; Robert Herman, Siamak A. Ardekani, and Jesse H. Ausubel, "Dematerialization," *Technology and Environment* (Washington, DC: National Academy Press, 1989), pp. 50-69. Dematerialization offers the possibility that economic growth may not inevitably lead to more materials use.

[29] Jesse H. Ausubel, "Regularities in Technological Development: An Environmental View," *Technology and the Environment*, Jesse H. Ausubel and Hedy E. Sladovich (eds.) (Washington, DC: National Academy Press, 1989), pp. 70-91.

Figure 2-5—Materials Used for Manufacturing Telecommunications Cables by AT&T Technologies

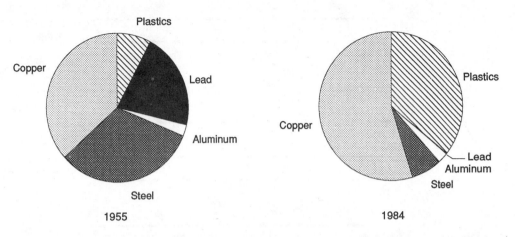

The composition of telecommunications cables illustrates the changing use of materials. Polyethylene has replaced lead as the dominant material in the cables' sheathing. This shows how material substitution can reduce the use of materials with adverse environmental impacts.
SOURCE: P.L. Key and T.D. Schlabach, "Metals Demand in Telecommunications," *Materials and Society* 10(3):433-451, 1986.

copper can be replaced by 65 pounds of fiber-optic cable, with the production of fiber consuming only 5 percent of the energy required for copper.[30]

Increasing Complexity

Statistics concerning materials consumption do not capture a more subtle change with potentially important environmental consequences: a trend toward increasing complexity of materials use. As noted earlier, advances in chemistry, materials science, and joining technology have made it possible to combine materials in new ways (e.g., anticorrosion coatings on metals, or fiber-reinforced composites) to meet performance specifications more cheaply. This creates products that are more complex from a materials point of view.

As an illustration, consider the modern snack-chip bag depicted schematically in figure 2-6. The combination of extremely thin layers of several different materials produces a lightweight package that meets a variety of needs (e.g., preserving freshness, indicating tampering, and providing product information).[31] The use of so many materials effectively inhibits recycling. On the other hand, the

package has waste prevention attributes; it is much lighter than an equivalent package made of a single material and provides a longer shelf life, resulting in less food waste.

Other products exhibit similar complexity. For example, automobiles are composed of a vast array of different materials, including high-strength steel, aluminum, copper, ceramics, metal-matrix composites, and more than 20 different types of plastic.[32] Even household laundry detergents contain many different components, each with a specific function: enzymes to dissolve biological stains, bleaches to whiten cleaned garments, and "builders" to prevent dislodged dirt from resettling on fabrics.[33] The greater complexity of these products offers benefits to consumers, but this complexity also makes it more difficult to evaluate their environmental attributes.

Driving Factors

While these trends have important implications for future environmental policy, they are evolving independently of environmental considerations. Instead, they are driven by economic factors, by

[30] U. Colombo, Proceedings of the Sixth Convocation of the Council of Academies of Engineering and Technological Sciences, pp. 26-27, 1988, cited in Herman, Ardekani, and Ausubel, op. cit., footnote 28.

[31] Council on Plastics and Packaging in the Environment, *COPPE Info Backgrounder*, "The Search for the Perfect Package: Packaging Design and Development," March 1992.

[32] Frank R. Field, *Materials Technology: Automobile Design and the Environment*, contractor report prepared for the Office of Technology Assessment, May 6, 1991.

[33] Andy Coghlan, "It May Be Green But Is It Clean?" *New Scientist*, May 4, 1991, p. 22.

Figure 2-6—Cross-Section of a Snack Chip Bag

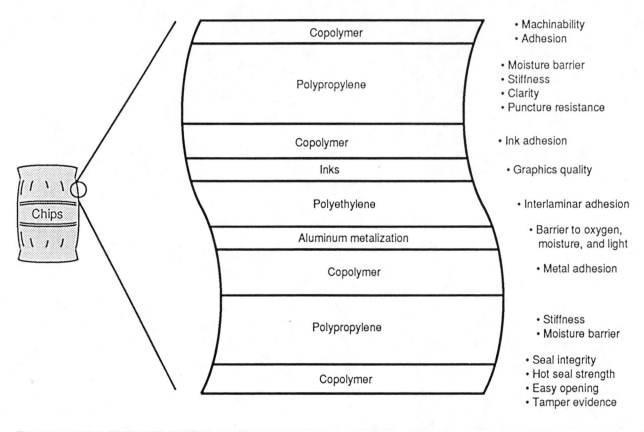

This cross-section of a snack chip bag illustrates the complexity of modern packaging. The bag is approximately 0.002 inches thick, and consists of nine different layers, each with a specific function. While such complexity can inhibit recycling efforts, it also can reduce the overall weight of the bag, and keep food fresher, thus providing waste prevention benefits.

SOURCE: Council on Plastics and Packaging in the Environment.

advances in technology, and by competition to satisfy changing consumer needs. For example, high repair costs prompt many customers to buy new goods rather than repair old ones.[34] This has encouraged the design of more sophisticated, self-contained products (e.g., consumer electronics with batteries sealed inside) that are intended to be used and thrown away. The creation of nonrepairable, nonserviceable items has also been motivated by liability concerns relating to product safety.[35] Meanwhile, improved manufacturing technologies have brought down the cost of such products, and more consumers can afford to purchase them, resulting in a greater number of goods that are discarded.[36]

The application of information technology to all stages of the production and marketing process has made shorter production runs affordable, enabling manufacturers to differentiate their product offerings and aim at narrower market niches. This has resulted in a proliferation of product lines (e.g., in automobiles, appliances, and computers). The increased ability to tailor products to individual needs comes at a time when changing lifestyles and a more diverse population are fueling demand for a wider

---

[34] ChemCycle Corporation, *Environmentally Sound Product Development in the Consumer Electronics and Household Battery Industries*, contractor paper prepared for the Office of Technology Assessment, July 1991.

[35] As goods have become more complex, the potential for consumer injury during repair has in many cases increased. It is not surprising then, that manufacturers are designing products so as to discourage consumer repair. Ibid.

[36] Ibid.

range of goods and services.[37] Elaborate production networks have been developed to meet the demands of these diverse markets (see chapter 4).

ENVIRONMENTAL POLICY IMPLICATIONS OF MATERIALS TRENDS

Are these trends good or bad for the environment? The answer is not always clear. Although complex products may be less recyclable, they may at the same time be more efficient in their use of energy and materials. For instance, advanced composite materials allow lighter components to be used in cars and aircraft, and consequently can lead to significant lifetime fuel savings and reduced air emissions. Multilayer food packaging can extend food shelf life. Steel-belted radial tires can be used year round, and are more durable than previous generations of tires.

With technologies available to create new materials and to combine conventional materials in new ways, designers are faced with more choices than ever before. Increasingly, these choices involve environmental dilemmas. Energy-efficient compact fluorescent bulbs, for example, contain mercury, a toxic heavy metal. In cases such as this, tradeoffs will be required, not only between traditional design objectives and environmental objectives, but among environmental objectives themselves: for example, waste prevention vs. recyclability, or energy efficiency vs. toxicity. In general, every design will have its own set of environmental pluses and minuses. This suggests several conclusions:

- The environmental evaluation of a product or design should not be based on a single attribute, such as recyclability; rather, some balancing of pluses and minuses will be required over its entire life cycle.
- The trend toward complexity seems certain to make the environmental evaluation of products more difficult and expensive in the future.
- Policies to encourage green design should be flexible enough to accommodate the rapid pace of technological change and a broad array of design choices and tradeoffs.

Looking Ahead

The pace of materials technology innovation continues to accelerate. As this chapter has shown, these innovations can lead to greater efficiency in materials use and less waste generated, measured per unit of production. This is an environmental triumph of a sort, since it means that environmental quality is greater than it would have been had these innovations not occurred. Policymakers should encourage these changes, but they must also recognize that what matters for future environmental quality is not just industrial efficiency, but the absolute quantity of resources used and wastes released to the environment. In absolute terms, more goods and services are being produced, and more wastes are being generated every year.

It is an open question whether present policies regarding economic growth can avoid irreversible environmental impacts or whether a drastic change is required. Conventional economic indicators do not address issues such as species loss and global climate change. To effectively face these challenges, measures of economic progress will have to be broadened to include not only industrial efficiency, but the overall health of human populations and ecosystems.[38] The next chapter explores how designers can begin to address these issues.

---

[37] More women working outside the home and less leisure time translates into increased demand for convenience products such as single serving packages and microwaveable dinners. See U.S. Congress, Office of Technology Assessment, *Technology and the American Economic Transition: Choices for the Future*, OTA-TET-283 (Washington, DC: U.S. Government Printing Office, May 1988), p. 22.

[38] There is a variety of ongoing work in this area. See Robert Repetto, ''Accounting for Environmental Assets,'' *Scientific American*, June 1992, pp. 94-100.

Chapter 3

Product Design and the Environment

Contents

Product Design and the Environment

Product design is a process of synthesis in which product attributes such as cost, performance, manufacturability, safety, and consumer appeal are considered together.[1] These principal design parameters are often constrained by regulatory requirements—for example, fuel efficiency targets, building codes, or tamper-proof packaging specifications. Thus, in virtually all cases, designers are forced to make tradeoffs among competing criteria.[2] At each stage of the design cycle, solutions are evaluated and reevaluated in light of a diverse ensemble of technical, economic, and social objectives. (For a discussion of how the design process works in the automotive industry, see appendix 3-A.)

The National Research Council has estimated that 70 percent or more of the costs of product development, manufacture, and use are determined during the initial design stages.[3] Design is therefore a critical determinant of a manufacturer's competitiveness. Because of the strategic importance of design, many corporations are adopting comprehensive programs for developing and introducing products.[4] With greater attention being given to the design process, new approaches to product development are emerging.

Companies are discovering that they cannot afford to have designers develop a concept in isolation and then toss it "over the wall" to production engineers. Instead, a "concurrent" design process is increasingly used, as depicted in Figure 3-1.[5] The product evolves continuously through a spiral of design, manufacturing, and marketing decisions. As a product progresses along the "design helix" toward commercialization, multidisciplinary product development teams take part in every major design iteration. This multifunctional approach safeguards product integrity and expedites product development from stage to stage. Implementation of concurrent design methods have allowed many firms to dramatically cut product cycle times, while delivering goods of superior performance and quality.[6]

The changing nature of design provides new opportunities for integrating environmental concerns into the product development process. The concurrent design methodology, with its multidimensional orientation, lends itself to the consideration of environmental impacts at every decision point. Similarly, total quality management (TQM) programs, which stress that quality must be "designed in," rather than tested for at the end of the production process, allow for a natural extension to

---

[1] Historically, design has been divided into the fields of engineering design and industrial design. Engineering design primarily specifies a product's technical characteristics, while industrial design is principally concerned with the "feel" of a product, such as styling and ease of use (ergonomics). Most products embody in varying degrees the inputs of these two disciplines. As used here, the term "designers" refers to all decisionmakers who participate in the early stages of product development. This includes a wide variety of disciplines: industrial designers, engineering designers, manufacturing engineers, graphic and packaging designers, as well as managers and marketing professionals.

[2] For a general discussion of the design process, see Nam Suh, *The Principles of Design* (New York, NY: Oxford University Press, 1990).

[3] National Research Council, *Improving Engineering Design: Designing for Competitive Advantage* (Washington, DC: National Academy Press, 1991), p. 1.

[4] For instance, Hewlett Packard, AT&T, and Ford have adopted such extensive product development strategies, sometimes known as "product realization" programs. Ibid.

[5] See, e.g., "Concurrent Engineering," *IEEE Spectrum*, July 1991, p. 22.

[6] Using concurrent planning techniques, Siemans Automotive has achieved extraordinary improvements in both productivity and quality. In 1975, Siemans produced 30,000 fuel injectors a month. In 1991, the company manufactured 30,000 fuel injectors a day with defect levels of 20 parts per million (.002 percent). Through the collaboration of designers and process engineers, the number of grinding steps was reduced six-fold. Over that same period, the direct human labor required for each fuel injector was reduced from 13 minutes to less than 2 minutes. Similarly, Motorola Inc., at one time required 30 days to build a pager. By implementing cross-functional design techniques and introducing significant levels of automation, a single pager can now be manufactured in 30 minutes. PBS Series: "Quality or Else! Challenge and Change," Oct. 18, 1991.

Figure 3-1—The Design Helix

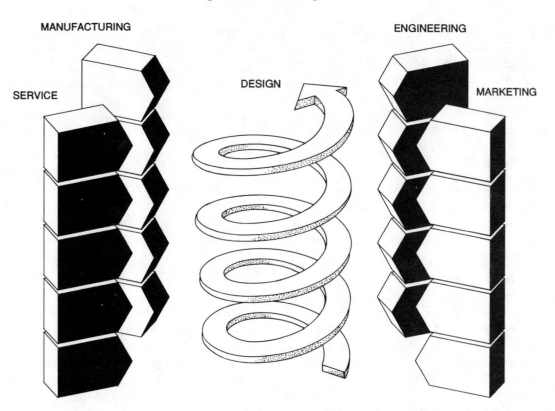

In a concurrent design process, each product discipline provides input into major design decisions. The interchange between disciplines reduces the time required for product commercialization.

SOURCE: GVO Design, Inc., Palo Alto, CA.

designing in the product's "environmental quality."[7][8]

WHAT IS GREEN DESIGN?

In general, products today are designed without regard for their overall impact on the environment. Nevertheless, many health and environmental laws passed by Congress do influence the environmental attributes of products. Some, such as the Clean Air Act, Clean Water Act, and Resource Conservation and Recovery Act, do so indirectly, by raising

industry's costs of releasing wastes to the air, water, and land. Others, such as the Toxic Substances Control Act and the Federal Insecticide, Fungicide, and Rodenticide Act, control the use of hazardous chemicals and pesticides directly.

Government regulations typically influence the design process by imposing external constraints, for example, requirements that automobile manufacturers comply with Corporate Average Fuel Economy (CAFE) standards, and with auto emissions standards under the Clean Air Act. The Office of

---

[7] See, e.g.: Global Environmental Management Initiative, *Proceedings of the First Conference on Corporate Quality/Environmental Management*, Washington, DC, Jan. 9-10, 1991; Charles M. Overby, "QFD and Taguchi for the Entire Life Cycle," ASQC Quality Congress Transactions, Milwaukee, WI, 1991; and W. David Stephenson, "Environmentalism's Strategic Advantage," *Quality*, November 1991, p. 20.

[8] Contemporary designers have available an array of tools that can simultaneously improve product quality while reducing environmental impacts. The use of computer-aided design and manufacturing tools can result in more effective utilization of materials—for example, Levi Strauss is using computers to test out new fabrics, patterns, and designs before ever cutting a piece of cloth. The use of just-in-time delivery methods optimizes inventory flows, and the integration of suppliers into the product development process ensures low defect levels and greater compatibility of product subcomponents. Finally, statistical quality control methods that identify process defects can improve factory efficiency and promote pollution prevention.

Figure 3-2—The Dual Goals of Green Design

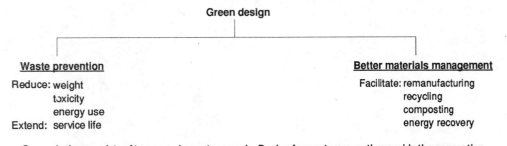

Green design consists of two complementary goals. Design for waste prevention avoids the generation of waste in the first place; design for better materials management facilitates the handling of products at the end of their service life.

SOURCE: Office of Technology Assessment, 1992.

Technology Assessment (OTA) uses the phrase ''green design'' to mean something qualitatively different: a design process in which environmental attributes are treated as *design objectives* or *design opportunities*, rather than as *constraints*. A key point is that green design incorporates environmental objectives with minimum loss to product performance, useful life, or functionality.

In OTA's formulation, green design involves two general goals: *waste prevention* and *better materials management* (Figure 3-2).[9] Waste prevention refers to activities by manufacturers and consumers that avoid the generation of waste in the first place.[10] Better materials management involves coordinating the design of products with remanufacturing operations or waste management methods so that after products have reached the end of their service life, their components or materials may be recovered and reused in their highest value-added application.[11] These goals should be viewed as complementary: while designers may reduce the quantity of resources used and wastes generated, products and waste streams will still exist and have to be managed.

Design for Waste Prevention

The old dictum that ''an ounce of prevention is worth a pound of cure'' is finding new relevance as industries attempt to modify traditional design and manufacturing practices. Examples of design for waste prevention include reducing the use of toxic materials, increasing energy efficiency, using less material to perform the same function, or designing products so that they have a longer useful life.

When a designer specifies a smaller quantity of a material, that decision has a multiplier effect on both the industrial and post-consumer waste streams.[12] Waste discharges, emissions, and energy consumed at each stage of the materials life cycle will decrease in proportion to the amount of material used (see box 3-A). Similarly, increasing the lifetime of products can result in direct waste reduction. Over a given time interval, less waste is generated during materials extraction, product manufacturing, and disposal. Related energy costs associated with processing and transport are also reduced.[13]

Product life extension can be achieved through use of more durable materials or through modular

[9] This formulation first appeared in U.S. Congress, Office of Technology Assessment, *Facing America's Trash: What Next for Municipal Solid Waste*, OTA-O-424 (Washington, DC: U.S. Government Printing Office, October 1989).

[10] See U.S. Congress, Office of Technology Assessment, *Serious Reduction of Hazardous Waste: For Pollution Prevention and Industrial Efficiency*, OTA-ITE-317 (Washington, DC: U.S. Government Printing Office, September 1986).

[11] The dividing line between waste prevention and better materials management is not always sharp. For instance, remanufacturing helps to conserve resources, and to avoid the generation of wastes that would otherwise have occurred. But OTA believes the distinction is nevertheless important to make. Waste management technologies generate environmental risks in their own right; by designing for waste prevention, these risks can be avoided.

[12] For example, for every ton of copper extracted in open-pit mining, 550 tons of materials are moved and processed. Mining and processing wastes include substantial emissions of arsenic, sulphur dioxide, and other byproducts. These wastes could be drastically reduced if copper was used more efficiently. See Robert Ayres, ''Toxic Heavy Metals: Materials Cycle Optimization,'' *Proceedings of the National Academy of Sciences*, vol. 89, No. 3, February 1992.

[13] Walter Stahel, ''Design as an Environmental Strategy,'' Paper presented at the Industrial Designers Society of America National Conference, Santa Barbara, CA, Aug. 8-11, 1990.

Box 3-A—*Getting the Lead Out*

The General Motors Delco Remy Battery Division has made significant strides in reducing hazardous constituents in both its products and processing operations. In 1974, a typical battery contained about 30 pounds of lead, whereas today a battery with much improved performance weighs only 19.5 pounds. This resulted in over 6 million pounds of lead waste prevention during 1990. In addition, the reformulation of alloy materials, changing from antimony-arsenic to calcium-tin, eliminated over 1 million pounds of antimony and arsenic waste in that same year.

Pollution prevention strategies have also involved an increased emphasis on in-process recycling. In one facility, 4.2 million pounds of lead, 730,000 gallons of sulfuric acid, and 250,000 pounds of polypropylene were reclaimed and reused. A new wastewater treatment process increases the percentage of lead in the resulting solids. This allows the lead to be more readily recycled. The solid precipitates are sent to a secondary lead smelter rather than a hazardous waste landfill.

SOURCE: GM Delco Remy Division.

Photo credit: Office of Technology Assessment

Some products can be redesigned to reduce the use of toxic substances. Over the past 5 years, manufacturers have reduced the level of mercury in household batteries by more than 85 percent.

designs that facilitate repair or upgrading of product components (see box 3-B). Products that are designed in a modular fashion have components of definable functionality that can be easily replaced or upgraded without affecting other components. This permits both products and product subcomponents to be easily serviced or refurbished.[14] It also allows product performance to be maintained over a longer time period, thereby obviating the need for buying entire new systems.[15]

However, the actual useful life of a product is affected by a number of external factors including maintenance practices, conditions of use, and the rate of technical or stylistic obsolescence.[16] While a number of industries have improved the durability of their products in recent years, a large percentage of materials that are extracted and processed through the economic system are still transformed into waste almost immediately.[17]

The belief that companies cynically pursue strategies of planned obsolescence in order to maximize profits is overstated.[18] Companies do shape consumer demand through their marketing strategies, but they also respond to customer demand for convenience and ease of product use. Since many consumers exhibit a greater sensitivity to a product's initial cost rather than its lifetime costs, this can inhibit the design of more durable, but expensive products. This sensitivity to cost is particularly evident in the area of energy-efficient home appliances and equipment—for example, air condition-

---

[14] Designing products so that they can be serviced is not mutually exclusive with designing for reliability. But due to proficient manufacturing methods and high labor costs, many complex products are designed to be extremely reliable over a given time period and then disposed (e.g., consumer electronics goods).

[15] For example, "modular upgradability" is quickly becoming a de facto standard in the personal computer industry. Fast growth companies such as Dell Computer Corp. and AST Research Inc. have based their success on designing modular machines. The designs permit customers to take advantage of the latest advances in microprocessor and memory technologies without buying a new computer. See *Wall Street Journal*, Sept. 10, 1991, p. B1, and *Electronic Engineering Times*, Oct. 28, 1991, p. 1.

[16] For example, steel-belted tires have twice the durability of tires that were made 20 years ago. If maintained properly, modern radial tires can last 60,000 to 80,000 miles. In practice, however, consumer misuse and neglect results in the tires wearing out much sooner. See "The Bumpy Road to Tire Recycling in America," *Garbage*, May/June 1991, p. 37.

[17] Robert Ayers, *Technology and Environment* (Washington, DC: National Academy of Engineering, 1990), p. 26.

[18] T. Teitenberg, *Environmental and Natural Resource Economics* (Glenview, IL: Scott, Foresman and Co., 1988), p. 191.

Box 3-B—Modular Design in the Housing Industry

An example of modular design is the Integrated Building and Construction Companies (IBACoS) project—a consortium of companies led by GE Plastics that have joined together to explore design concepts for homes of the future. In a demonstration house built in Pittsfield, Massachusetts, several key design ideas have been implemented.

One principal concept is the notion of designing a home as a collection of "disentangled" systems—systems that can be developed, produced, and installed independently of one another. Heating, lighting, power, and plumbing systems are designed to permit adaptability and flexibility to accommodate changing lifestyles. This flexible design strategy allows various elements of these systems to be upgraded with the latest technologies. Thus, the energy efficiency of a home could be continually improved, and critical components can be accessed without requiring destruction of walls or floor/ceiling structures. For example, kitchens or bathrooms could be enlarged, modified, or moved to different parts of the home. Also, as new information services become available to households, fiber-optic cable could be brought into a home without displacing existing wiring or fixtures.

The main thrust of the IBACoS strategy is to develop a set of core systems—a kit of parts designed and coordinated by computer, manufactured by member companies, and capable of being used in a wide variety of ways in both new and old buildings. This approach envisions significant cooperation among many different players. Consortium participants will include architects, builders, materials suppliers, and an

Photo credit: GE Plastics

General Electric Plastics has constructed a 2,900 square-foot home that is designed to provide a "living laboratory" for the development of advanced building systems and components. The home is made from recycled engineering thermoplastics and a variety of traditional construction materials.

array of manufacturers with expertise in pre-built structural components, energy and electrical systems, and kitchen and bath systems. Streamlined delivery and storage of "core systems" will provide quick response capability to meet customer needs. The IBACoS network will thus be able to provide a diverse set of product offerings while fostering resource efficiency and environmental quality.

Houses represent one of the largest potential markets for secondary materials. In the IBACoS scheme, materials efficiency would be encouraged by using recycled plastics, high strength composite woods made from wood scrap, and reconstituted gypsum. For instance, plastic roof shingles have been produced from discarded computer equipment. In addition, by using prefabricated structural components and subsystems, the generation of on-site waste and scrap could be reduced.

SOURCE: Michael Dickens, Director-IBACoS Program.

ers, refrigerators, and light bulbs. Consumers usually do not invest in energy efficiency unless it offers a fairly short payback—typically less than 2 years for home appliances.[19]

Design for Better Materials Management

By and large, resources flow through our society in one direction only. Designers rarely think about how their products will be managed as wastes after their useful life is over. And waste management providers tend to accept the composition of waste streams as a given. If product design and waste management were coupled more closely, this could reduce the cost of materials to industry and address environmental problems at the same time. This will require coordinated research on both principles of design and improved waste management processes.

[19] See U.S. Congress, Office of Technology Assessment, *Building Energy Efficiency*, OTA-E-518 (Washington, DC: U.S. Government Printing Office, May 1992).

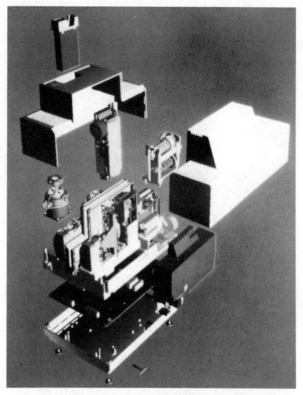

Photo credit: NCR Canada Ltd.

The NCR 7731 Personal Image Processor features the
latest in optical imaging technology and incorporates a
number of green design concepts. The product features
modular components that can be readily disassembled,
parts consolidation, and the use of recycled materials.
The modular design allows the device to be configured
to meet specific customer needs, and so avoids
unnecessary hardware and reduces waste.

Table 3-1—Principles of Design for Disassembly

Minimize material variety
Use compatible materials
Consolidate parts
Reduce number of assembly operations
Simplify and standardize component fits and interfaces
Identify separation points between parts
Use water-soluble adhesives when possible
Mark materials to enhance separation

SOURCE: General Electric Plastics, Pittsfield, MA.

tions until the plastic is finally incinerated to recover
the chemical energy.[20]

Design for Remanufacturing and Recycling

Giving consideration to how product components
or materials can be reclaimed will likely cause
companies to alter conventional design and manufacturing strategies.[21] Although not widely prac-
ticed, design for remanufacturing can be attractive
from both an environmental and a business point of
view (see box 3-C). Similarly, recycling offers a
number of potential benefits. Recycling can reduce
virgin material extraction rates, wastes generated
from raw material separation and processing, and
energy use associated with manufacturing. It can
also divert residual materials from the municipal
waste stream, relieving pressure on overburdened
landfills.

Products that can be rapidly disassembled into
their component parts lend themselves both to
remanufacturing and recycling (see table 3-1). Design for disassembly can go a long way toward
establishing both closed-loop production-reclama-
tion systems where components and materials are
reused in the same products, and open-loop systems
where materials are recycled several times for use in
different products. A number of durable products
including automobiles, refrigerators, and cooking
appliances are beginning to embody aspects of this
design approach.[22] However, durable products present special problems because it is difficult for
designers to anticipate how waste management

Examples of design for better materials management include making products that can be remanufactured, recycled, composted, or safely incinerated
with energy recovery. Broadly speaking, these
management options are listed in order of preference, both from a business perspective and an
environmental perspective. One model of plastics
management, for instance, envisions a life cycle in
which virgin plastic components are reused as long
as possible, then the materials are repeatedly recycled through lower and lower value-added applica-

---

[20] This model of materials management has been proposed by GE Plastics, Pittsfield, MA.

[21] See Michael E. Henstock, "The Conflict Between First Cost and Recyclability in the Design of Manufactured Goods," *Resources Policy*,
September 1978, pp. 160-165; "Design for Recycling,"Institute of Scrap Recycling Industries, *Phoenix Quarterly*, vol. 21, No. 1, winter 1989; and Rick
Noller, "Environmentally Responsible Product Design," paper presented at the National Academy of Engineering Workshop on Engineering Our Way
Out of the Dump, Woods Hole, MA, July 1-3, 1991.

[22] "Built to Last—Until It's Time To Take It Apart," *Business Week*, Sept. 17, 1990, pp. 102-106.

Box 3-C—Remanufacturing

When durable goods such as kitchen appliances or machine tools wear out, they are usually discarded. But there exists another option that may offer considerable economic and environmental benefits: remanufacturing. Remanufacturing involves the restoration of old products by refurbishing usable parts and introducing new components where necessary. It simultaneously results in product life-extension (a form of waste prevention) and promotes reuse of subcomponents and materials. Thus, in the case of remanufacturing, waste prevention and materials management strategies can be mutually reinforcing.

Because of the economic advantages that can accrue from remanufacturing, a variety of different industries are embracing the concept. For example, Xerox Corp. restores and remanufactures many used parts from its copiers, including electric motors, power supplies, photo-receptors, and aluminum drums. Xerox is now recycling about 1 million parts a year in this way, resulting in savings around $200 million. The parts are used as both replacement components and in new equipment. To facilitate the refurbishing and recycling of various components and product subsystems, Xerox is standardizing its designs so that a larger number of parts can be used in a variety of different products. The company has set up its remanufacturing lines in parallel with its new production lines to achieve the same levels of high quality. It has also involved its suppliers more directly in the design process, so that opportunities to use recycled components and materials, especially plastics, will not be overlooked.

The use of replacement parts for automobiles and trucks is one of the most prevalent applications of product remanufacturing. For instance, Arrow Automotive Industries, a company that remanufactures automotive components such as starter motors, clutches, and carburetors, has annual sales of approximately $100 million. However, the largest single remanufacturer in the United States is the Department of Defense. Military equipment and systems ranging from aircraft and radar to rifles are remanufactured on a regular basis to extend the life of expensive technological hardware.

Apart from the economic benefits that can accrue to a manufacturer, the reuse of high value-added components takes advantage of the original manufacturing investment in energy and materials. This yields greater environmental benefits than simply recycling the constituent materials of the components. In most cases, the energy embodied in a new product is many times that needed to remanufacture the same product.

SOURCES: Jack Azar, Xerox Corporation, personal communication, Aug. 15, 1991; Robert T. Lund, "Remanufacturing," *Technology Review*, February/March 1984, pp. 19-29.

---

practices might change by the time the product enters the waste stream.[23]

Just as important as designing for materials recovery is designing for the *use* of recovered materials. Developing design configurations that facilitate the disassembly and separation of product components is not enough. Companies must actually incorporate recycled materials and components into their products to bring about true environmental benefits. While the primary barriers to recycling are economic,[24] the limited availability of high-quality recovered materials can also complicate efforts to introduce recycled materials into new designs.[25] Contamination and indiscriminate mixing of materials during collection and separation can undermine recycling efforts, and chemicals added in the original manufacturing process may be difficult to remove, or may degrade the properties of reprocessed materials.[26] Even if materials are free of

---

[23] For example, even if designers radically altered the design of automobiles and household appliances today, current models would continue to enter the waste stream well into the next decade. Other products, such as household chemicals and batteries, can linger in basements and garages for years until eventual disposal. Thus, such time lags can complicate the efforts of designers to incorporate environmental objectives into their designs.

[24] William L. Kovacs, "Dark Clouds Carry Silver Linings: Recyclable Materials Are the Basis for a Competitive Industrial Policy," *Resource Recovery*, August 1989, pp. 5-6.

[25] The problems associated with the collection and processing of recycled materials are discussed extensively in the OTA report, *Facing America's Trash*, op. cit., footnote 9, pp. 135-190.

[26] As an illustration, glass cullet itself is 100 percent recyclable, but it is difficult to make glass entirely from cullet because cullet lacks "fining" agents that are needed to reduce bubbles in the glass. (See Testimony of the Glass Packaging Institute before the Subcommittee on Environmental Protection of the Senate Committee on Environment and Public Works, June 6, 1991; Also see OTA, *Facing America's Trash*, op. cit., p. 151). In the case of aluminum, the presence of mixed alloys in discarded aluminum goods complicates the secondary production process. Unless the alloy mix is controlled precisely, the recovered aluminum will fail to meet product specifications. (See R.E. Sanders and A.B. Trageser, "Recycling of Lightweight Aluminum Containers: Present and Future Perspectives," Proceedings of the Second International Symposium-Recycling of Metals and Engineered Materials, held by the Minerals, Metals & Materials Society, October 1990).

Photo credit: Xerox Corp.

Xerox reuses and remanufactures many of the sophisticated components from its copiers. Remanufactured machines and new machines are assembled on the same production line to the same quality standards. **Top left:** "Remanufactured" and "new build" assemblies on the same cart adjacent to the production line. **Above:** In the middle of the production line the two machines appear similar—"remanufactured" on the left and "new build" on the right. **Bottom left:** At the end of the production line the two machines are indistinguishable.

external contamination, recycling processes may degrade the materials; for instance, paper fibers degrade with each successive reuse.

If designers are to use recovered materials more extensively, they must have confidence that these materials can provide similar performance and properties as virgin materials. This may be best achieved if their accustomed materials suppliers offer recovered materials with guaranteed properties alongside their offerings of virgin materials.[27]

Design for Composting and Incineration

Apart from recycling and remanufacturing, there are two other materials management options that designers can consider: composting and incineration.[28] Designers can facilitate composting by making products entirely out of biodegradable materials.[29] For example, starch-based polymers and films can substitute for plastic in a variety of applications.[30] These starch-based polymers are inherently biodegradable, and easily composted.[31] Similarly, products could be designed for safe

---

[27] Charles Burnette, Chairman, Industrial Design Department, The University of the Arts, Philadelphia, PA, personal communication, Sept. 1, 1992.

[28] Composting refers to the process of biological decomposition of solid organic materials by microorganisms (mainly bacteria and fungi). "Compost" is the stabilized, humus or soil like product of this process.

[29] As an illustration, Procter & Gamble's *Pampers* and *Luvs* brands contain about 80 percent compostable material. But the plastic backsheets on the diapers are not compostable. Thus, the compostable material must be separated from the backsheet before composting. To eliminate this separation stage, P&G is currently developing backsheets made from compostable material. See the comments of Edward L. Artzt, Chief Executive Officer, Proctor & Gamble in *Beyond Compliance: A New Industry View of the Environment*, Bruce Smart (ed.) (Washington, DC: World Resources Institute, April 1992), pp. 36-40.

[30] One company, Warner-Lambert, recently opened a large-scale manufacturing facility to produce such agriculturally derived polymers. The trade name of the polymer is Novon.

[31] See Ramani Narayan, "Bioremediation/Biodegradation of Plastic Wastes by Composting," Proceedings of the Global Pollution Prevention Conference and Exhibition, Washington, DC, Apr. 3-5, 1991.

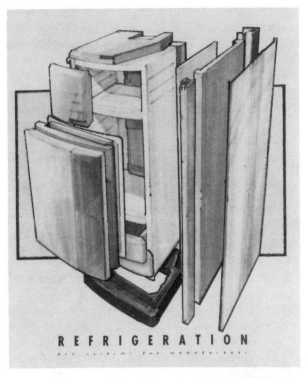

Photo credit: GE Plastics

The modern-day refrigerator is typically not designed for recyclability, and consequently is almost impossible to disassemble. Refrigerators use large amounts of polyurethane foam (this foam contains CFCs) that cannot be easily separated from different metal components. New refrigerator designs are beginning to incorporate modular concepts, as well as alternative forms of insulation such as silica aerogels or vacuum-based insulation.

incineration by avoiding the use of heavy metals and chlorinated organics.

For these opportunities to be realized, though, requires that product design changes be coordinated with new *systems* of product disposal and integrated

waste management (see ch. 4). If products are designed for composting or safe incineration, but end up being landfilled, the design improvements are effectively nullified.[32] Historically, there has been little, if any, coordination between the stages of design and waste management. This situation needs to change if society is to benefit from the environmental leverage afforded by design. Promoting greater coupling between manufacturing and waste management is a major challenge for policymakers (see ch. 6).

HISTORY OF GREEN DESIGN

The idea of green product design is not new. It was developed in the late 1960s and early 1970s, along with the explosion of environmental consciousness that led to the creation of the Environmental Protection Agency and to the passage of laws such as the Clean Air Act, Clean Water Act, and the Resource Conservation and Recovery Act.[33] During the 1980s, ideas such as design for remanufacturing and design for recycling were developed in technical journals and conferences, but the concept did not receive much attention from policymakers or the public.[34] Perhaps because of recent alarming headlines about global climate change, ozone depletion, and overflowing landfills, the issue has enjoyed a renaissance in the past few years. Several recent books and articles have explored how architects, engineers, industrial designers, packaging designers, and graphic designers can incorporate environmental attributes into their designs.[35]

Despite this 20 year history, however, the concept of green design has not yet been integrated into engineering education or practice. Indeed, until recently, "design for the environment" meant a design that protects the product against the effects of

---

[32] Approximately half of the material currently disposed in landfills is potentially compostable. But even the most readily compostable portions of the waste stream, like yard clippings, are rarely composted because of poor public education and inadequate waste management. (See OTA, *Facing America's Trash,* op. cit., footnote 9.)

[33] The idea was already well developed in 1970, the earliest year of OTA's literature search. See Jacob Friedlander and Merril Eisenbud, "Environmental Dangers Challenge Design Engineers," *Mechanical Engineering,* November 1970, p. 15. A discussion of environmentally sound product design and policy options to encourage it appears in the Second Report to Congress, "Resource Recovery and Source Reduction," U.S. Environmental Protection Agency, 1974. See especially appendix B: "Product Design Modifications for Resource Recovery, Source Reduction, or Solid Waste Purposes."

[34] Charles Overby, "Product Design for Recyclability and Life Extension," American Society for Engineering Education Annual Conference Proceedings, 1979, p. 181; Robert T. Lund, "Remanufacturing," *Technology Review,* February/March 1984, p. 19.

[35] See Avril Fox and Robin Murrell, *Green Design* (London: Architecture Design and Technology Press, 1989); Dorothy Mackenzie, *Design for the Environment* (New York, NY: Rizzoli International Publications, Inc., 1991); David Wann, *Biologic: Environmental Protection by Design* (Johnson Books, 1990); Charles Overby, "Design for the Entire Life Cycle: A New Paradigm?" American Society for Engineering Education Annual Conference Proceedings, summer 1990, and references therein; The World Wildlife Fund and Conservation Foundation, "Getting at the Source: Strategies for Reducing Municipal Solid Waste," Washington, DC, 1991; Tedi Bish and Suzette Sherman, "Design To Save the World," *International Design,* November/December 1990, p. 49.

Photo credit: Office of Technology Assessment

Consumers sometimes can leave separated materials at igloos or other containers placed in conspicuous areas by communities or firms running recycling programs.

moisture, corrosion, or weather. Designers' use of materials have undergone dramatic changes over the past 50 years, but these changes have evolved independently of environmental concerns, being driven primarily by technological innovation and economic competition among materials (see ch. 2).

This situation is changing rapidly, however. Many companies, large and small, are starting to change their process and product designs in ways that reduce both their own waste disposal problems and those of their customers.[36] Several government-funded projects are underway in the United States and Europe to develop environmental handbooks or checklists for designers (see chs. 5 and 6). For example, researchers in the Netherlands have developed computer software to assist designers in making environmentally sound choices.[37]

ALTERNATIVE VIEWS OF GREEN DESIGN

The idea of green design seems straightforward, but there is no rigid formula or decision hierarchy for implementing it. One reason is that what is "green" depends strongly upon context. While there are some environmental design imperatives that are sufficiently compelling to apply to many different products (e.g., avoiding the use of chlorofluorocarbons), in general green choices will only become clear with respect to specific classes of products or production networks. What constitutes green design may depend on such factors as the length of product life, product performance, safety, and reliability; toxicity of constituents and available substitutes; specific waste management technologies; and the local conditions under which the product is used and disposed. For example, designing a product to be recyclable makes little difference if the infrastructure for collecting and recycling the product do not exist.

On a deeper level, though, one's philosophical view of the relationship between the economy and the environment strongly conditions one's view of green design and the environmental "problems" it should address. One taxonomy of this relationship employs a set of five paradigms, ranging from "frontier economics" to "deep ecology."[38] Here we will discuss alternative views of green design for the three intermediate paradigms: "environmental protection," "resource management," and "eco-development."

Paradigm 1: Environmental Protection

In this paradigm, the environment is recognized as an economic externality that must be safeguarded through laws and regulations. Tradeoffs are seen between industrial competitiveness and protecting the environment (e.g., employment vs. protecting endangered species), and cost-benefit analysis is offered as a means of balancing the two. This view is fundamentally anthropocentric, with the principal concern being the effect of pollution on human health and welfare.

The "problem" in this case is that human society produces too much waste. This concept leads to policies that focus on reducing the quantity or toxicity of waste: e.g., waste prevention, recycling, or treatment. Similarly, the objective of green design should be to reduce the quantity and toxicity of

---

[36] A diverse set of companies including 3M, Xerox, AT&T, Procter & Gamble, S.C. Johnson Wax, and Eveready have implemented programs of process and product reformulation to reduce levels of waste generation at both the manufacturer and post-consumer stages. See *Beyond Compliance: A New Industry View of the Environment,* op. cit., footnote 29.

[37] The program, called SimaPro, is available from PRé Consultants, Amersfoort, The Netherlands.

[38] One extreme, *frontier economics,* focuses on economic growth and emphasizes free markets and unbridled exploitation of resources. The other extreme, *deep ecology,* focuses on harmony with nature and emphasizes drastic reductions in human population and the scale of human economies. See Michael E. Colby, "Environmental Management in Development," World Bank Discussion Papers, Washington, DC, 1990.

wastes requiring disposal, e.g., making products more recyclable, light-weighting, etc. Progress is measured in terms of increasing the efficiency of energy and materials use, i.e., reducing the quantity of energy and materials required per unit of production. This view does not concern itself explicitly with whether the physical flows of energy and materials through the economy are ecologically "sustainable."

Paradigm 2: Resource Management

In this view, the environment is recognized as an economic externality that must be internalized in measures of economic performance and policy decisionmaking. The earth is seen as a closed economic system, and therefore the main challenge is to "economize ecology." If those who use resources and generate pollution are made to pay the true price of those environmental services, this will lead to sustainable industrial development. Advancing technology is seen as an integral part of achieving more efficient use of energy and materials. Technologically advanced countries should aggressively transfer new, more efficient technology to developing countries, and assist them in stabilizing their populations.

The "problem" in this paradigm is that human society is managing its resources poorly, generating pollution that threatens to undermine the ecological productivity upon which the economy depends. The solution is to "get the prices right" through taxes on resource use and pollution, or perhaps tradable permits to pollute within sustainable limits. Such economic incentives are seen as providing more flexibility than regulations, so that industry can respond in the most cost-effective way.

This view assumes that environmental services can be monetized, and that functioning markets for these services can be created. It does not address uncertainties in the valuation of these services or in the correct determination of the relevant ecological thresholds or carrying capacities. It is primarily anthropocentric, since it is concerned with the stock of "resources" available for human use, but extends its concern to quality of life of future generations as well as the present generation. Sustainable development is defined as maintaining a nondecreasing stock of human plus natural capital, implying some substitutability between the two.[39]

In the resource management paradigm, green design involves choices that conserve resources as well as reduce wastes. Emphasis is on the materials inputs in products, e.g., avoiding the use of materials that are toxic or become dispersed in the environment. In principle, the prices of material inputs would reflect their demand on environmental services, thus providing the correct signals to the designer. The resulting price changes would cause reorganization of the production system toward cleaner technologies and discarded materials would have a higher value, thus encouraging recovery and recycling.

Paradigm 3: Eco-Development

The eco-development paradigm stresses the co-evolution of human society and ecosystems on an equal basis. The earth is seen as a closed ecological system and therefore the principle challenge is to "ecologize the economy." This view is less anthropocentric than the resource management view, emphasizing that nature has an intrinsic value that is independent of the value placed upon it by the human economy. Thus, this view has a moral or ethical dimension that implies a transformation of societal attitudes toward nature (not assumed in the previous paradigms).

The "problem" in this case is that the scale of human economic growth is inconsistent with the long-term coexistence of man with nature. Sustainability is defined as nondecreasing stocks of human and natural capital maintained independently; that is, no substitutability between technology and natural resources is assumed.[40] In the face of uncertainty about ecological thresholds and the world's carrying capacity, the "precautionary principle" applies: new technologies or development projects must demonstrate that they are consistent with sustainability as defined above before they are adopted. Progress is measured not in terms of efficiency, but in terms of the health of regional ecosystems as well as human health.

Policy objectives for development under this paradigm include moving toward a closed materials

---

[39] This has been called the criterion of "weak sustainability." See Herman E. Daly and John B. Cobb, *For the Common Good: Redirecting the Economy Toward Community, the Environment, and a Sustainable Future* (Boston, MA: Beacon Press, 1989).

[40] This has been called the criterion of "strong sustainability." Ibid.

cycle. The economy would rely principally on renewable sources of energy and materials, extracted at rates that would not affect ecological health. Nonrenewable resources would be recovered and recycled indefinitely. Instead of tradable pollution permits, tradable permits might be issued for the extraction of a fixed quantity of nonrenewable materials.[41] The production/consumption system would be restructured to optimize the utilization of goods to satisfy essential human needs, rather than the ownership of goods to satisfy frivolous "wants." Green designs would avoid use of materials that are toxic to humans or ecological systems, substitute renewable for nonrenewable materials, and ensure that nonrenewable materials could be readily recovered for recycling.

Analysis

These three paradigms illustrate the different assumptions that underlie the environmental policy debate. They reflect different views of mankind's place in the natural world, and of its obligations to future generations as well as to other species. Present U.S. policy is most closely approximated by the environmental protection paradigm, while many environmental groups espouse the eco-development perspective. Resource management is the theme of reports such as the Brundtland Commission's "Our Common Future," the Worldwatch Institute's annual "State of the World," and the World Resources Institute's annual "World Resources."[42]

These paradigms also suggest different criteria for defining green design. In the environmental protection view, a product design may be considered green if it results in 10 percent less waste than last year's design over its entire life cycle (waste prevention). The same design may be rejected from the eco-development perspective because it uses nonrenewable materials that are not recycled and do not biodegrade. Evidently, green product design within

each succeeding paradigm involves satisfying a correspondingly broader set of criteria for compatibility with the natural environment.

In this chapter, we have defined green design as an extension of traditional design to include the goals of waste prevention and better materials management. This formulation might be criticized as being too conservative, since it suggests a narrow focus on the "outputs" of the production system that is characteristic of the environmental protection paradigm. Certainly, other formulations are possible. For example, an alternative definition focusing more on the "inputs" might involve reducing the use of toxic materials, and relying more on resources that are managed in a "sustainable" way. Such a definition might be more consistent with the eco-development paradigm.

In some cases, designers may have information about materials choices that bear directly on the destruction of irreplaceable resources, or the extinction of endangered species. An example might be avoiding the use of tropical hardwoods that are harvested from environmentally sensitive rainforests. In most cases, though, it will not be clear which choices are more ecologically "sustainable." It seems most practical to address global issues of ecological sustainability at the level of national policy, rather than at the level of the individual designer.

OTA chose a formulation of green design that suggests the most concrete actions available to the designer. A narrower focus on waste prevention and better materials management provides tangible criteria for evaluating the choices that designers make every day. The next chapter discusses various strategies that designers and companies can employ to reduce the environmental impacts of their products.

[41] See Herman E. Daly, *Economics, Ecology, and Ethics: Essays Toward a Steady State Economy* (San Francisco, CA: W.H. Freeman, 1980), pp. 337-348.

[42] Colby, op. cit., footnote 38.

APPENDIX 3-A:
THE AUTOMOBILE
DEVELOPMENT PROCESS[1]

Introduction

The automobile sold in the United States today is a complex product, not only in terms of the functions that it performs, but also in terms of the marketplace, and in terms of the wide set of goals, both private and public, that it is expected to meet. This appendix is intended to convey the many factors that influence automobile design decisions, and how environmental concerns enter the process. The following description of the automobile design process is necessarily generalized, but it captures the key issues and tradeoffs that govern contemporary automobile development.[2]

Automobile Product Development

Concept—The first stage in automobile design and development can be called "concept development." This stage in the process is essentially a strategic effort, which can take one of two forms. Most commonly, a particular set of market segments is identified, defined not only by demographics like age and marital status, but also by income, spending characteristics, and stylistic trends. The car concept that evolves from these considerations is a combination of appearance, features, and cost that is expected to attract enough purchases from the targeted groups to justify the development effort and to make the automaker money. As Charles Centivany, a Ford product planning manager, said, "In one sense we are looking for customer demand to pull us along, while looking for pockets not being filled, or that could be filled better."[3]

Another concept stratagem is to develop a product which can be used as a testbed for innovative vehicle technologies, either in manufacturing or in the product itself. The classic example of this kind of development in the domestic automobile industry has been the Chevrolet Corvette and, more recently, the Pontiac Fiero. Because the production volume is low, limited testing of innovative automobile tech-

nologies can be performed with low risk to the producer, and a wide range of innovations can be easily tested. For example, the composite automobile leaf spring was first introduced on the Corvette, although it now can be found in several other General Motors vehicles. Of late there has been considerable use of this stratagem at General Motors to develop manufacturing technologies. The Fiero introduced the space-frame vehicle manufacturing process, which has been considerably refined in two current General Motors products, the Saturn and the All Purpose Vehicles (APVs).

Whatever the original source of the ideas, the purpose of a vehicle concept is to supply an outline of the basic characteristics of the product under development, and a set of guidelines against which the results of the design process are to be measured.

Design Studio—The automobile concept is then passed to the design studios, where the concept is fleshed out on paper and, ultimately, in clay for review by the concept team and upper management. The focus at this phase of the process is to develop a vehicle shape that can accommodate the concept requirements while achieving those intangible characteristics known as "style." As a consequence, the studio draws upon a wide range of inputs in the course of developing the shape of the vehicle. These include past features of the product line as well as competitive product lines.

Although the design studio has historically drawn from U.S. sources, the recent globalization of the automotive market has led to international partnerships and outright acquisition of centers of styling excellence. In particular, the U.S. original equipment manufacturers (OEMs) have focused this effort on Italian and British design shops, although many elements of Japanese design have also been incorporated. Although the nation of origin of most designs can still be identified on sight, there has been a trend toward blurring the distinctions between the different schools of design. However, the Japanese have proven to be most mutable, as they have located many elements of their design effort in the United States, particularly California. For example, the

---

[1] This discussion is drawn from Frank Field, "Automobile Design and the Environment," contractor report prepared for the Office of Technology Assessment, May 1991.

[2] For a good overall view of the process, see James P. Womack, Daniel T. Jones, and Daniel Roos, *The Machine That Changed the World* (New York, NY: Rawson Associates, 1990).

[3] Christopher A. Sawyer, "It's All in the Planning," *Automotive Industries* (Radnor, PA: Chilton Co., January 1991), p.20.

popular Mazda Miata is the product of a U.S. design shop, and its appearance reflects these origins.

Frequently, the design studios will devise several potential vehicles for any one concept. These alternatives are winnowed down within the design studios and by corporate decisions until a single vehicle geometry is settled upon, usually following the presentation of a full-scale clay mock-up of the vehicle. Once the clay models have been approved by concept and the higher levels of management, car development is turned over to advance engineering.

Advance Engineering—Advance engineering is the stage in the vehicle development process where what most people think of as product design really happens. Here, the product of the design studios and the concept teams is converted into the first real engineering drawings of the automobile.

The classic approach to this problem is to divide the vehicle into functional subsystems, such as the body, the chassis, the powertrain, and the interior. The division of the vehicle into subsystems is a critical organizational simplification of the vehicle development process. These subsystems are defined to isolate engineering decisions within the subsystem design group. Without this isolation, the engineering problem is simply too large to be satisfactorily resolved. Instead, specific requirements (known as "design bogeys") are developed at the advance engineering level which must then be implemented by the product engineers. Provided that these bogeys accurately reflect the objectives devised at the concept level, and are satisfactorily backed up with good engineering practice, the automobile can be successfully designed.

Two of the most critical design bogeys established at advance engineering are cost and weight. Both of these are central to the success of the designers in meeting the requirements of the vehicle concept. Cost targets must be met in order to meet the pricing objectives that underlie the marketing strategy, and weight targets are critical to assuring that the vehicle performance (i.e., fuel economy and vehicle handling) will meet the concept goals.

Since 1978, fuel economy specifications have been principally dictated by Federal Corporate Average Fuel Economy (CAFE) requirements. No automaker can afford to ignore CAFE when devising its vehicle designs. CAFE enters into the automobile design cycle at its inception. The strategists, in the course of defining the vehicle concept and the product strategy, must establish a target fuel economy for the product. The design studios are not directly affected by this target, although aerodynamic drag and vehicle rolling resistance (two key factors along with engine performance that determine vehicle fuel economy) are a direct consequence of vehicle shape and weight, respectively. However, once the concept is passed to advanced engineering, the need to meet CAFE becomes one of the critical design parameters, probably second only to cost.[4]

Thus, weight bogeys become the primary way in which fuel economy is managed by the advance engineering departments. For a new design, the advance engineering groups will establish weight targets for each of the major vehicle subsystems. At the same time, the materials composing those subsystems are largely determined, particularly for the body and the chassis. CAFE has thus encouraged automakers to use more lightweight materials like plastics or plastics composites. This has raised a number of concerns about the recyclability of automobiles. With a decreasing metal content in cars, existing auto scrap dealers are finding it increasingly difficult to maintain business viability.[5]

Apart from CAFE requirements, designers must also give consideration to vehicle emission and safety regulations. The need to meet certain emission levels affects engine performance specifications,[6] while safety standards affect a number of design parameters including the choice of materials.

---

[4] CAFE regulations have a ripple effect all the way down to automobile suppliers. For example, Goodyear's new "environmental tire" is designed to improve fuel efficiency. The tire weighs less and has reduced rolling resistance.

[5] Automakers are working with the plastics industry to develop the technologies necessary to make the recovery of plastics economical, but difficult barriers remain to be overcome (see box 4-C, "Design and Materials Management in the Auto Industry").

[6] The aims of limiting emissions and improving fuel economy have a peculiar interaction. On one hand, improved fuel economy implies that energy is more efficiently extracted from the fuel. If so, a greater fraction of the available fuel is burned (reducing hydrocarbon emissions) and a larger fraction of the fuel is completely combusted (reducing the amount of carbon monoxide released). On the other hand, this improvement in efficiency is generally achieved by operating the engine at a higher temperature, which unfortunately increases the amount of nitrogen oxides that are produced. Additionally, changes in the operating condition of the engine (higher speeds, acceleration, etc.) require changes in the way in which combustion is affected (spark advance, timing, etc.).

Once the basic design bogeys are established, the advance engineering groups turn to developing the first engineering drawings of the vehicle subsystems. These drawings are fairly general, since much of the detail work requires more resources than are usually available at this level. However, it is at this stage in the design that the basic shapes of the critical vehicle elements are devised, and where the majority of the vehicle material specifications are made.

The most important element of engineering design at this and subsequent stages in the automobile design process is past experience. Vehicle designs almost always start with a consideration of past designs having similar requirements. Automobile designers rarely start from "blank paper" when designing vehicles, primarily because it is inefficient for them to do so. There are several reasons for this:

1. *Time pressure*: A crucial element of the automobile development process is the issue of time. Automakers have found that, like so many other industries, time to market is central to market competitiveness. While tooling acquisition and facilities planning are major obstacles to shortening the development cycle, they tend to be outside direct control of the automaker. Design time, however, is directly under the control of the automaker, and reduction of design time has been a major goal in vehicle development.
2. *Cost pressures*: The reuse of past designs also saves money. In addition to the obvious time savings described above, the use of a proven design means that the automaker has already developed the necessary manufacturing capability (either in-house or through purchasing channels). Furthermore, because the old part has a known performance history, the product liability risk and the warranty service risk are also much reduced.
3. *Knowledge limitations*: Underlying factors 1 and 2 is the fact that the automobile engineering design community is still developing the information and analysis base needed to do analytical design of automobile components. This limitation does not arise from a lack of engineering talent, although some of the domestic OEMs have had a tendency to lay off engineering staffs when times get hard. Rather, the limitation is a consequence of a real lack of knowledge of the structural loads that the

various automobile subsystems must be able to sustain. In other words, the automakers have only a rough idea of what loads a car will experience in service. Given this limitation, it is far more efficient to start with a past design which has proven to be successful, and to modify it to meet the geometric limitations of the new vehicle. Starting at this point, and backing up the design with prototyping and road testing, has proven to be far more efficient.

This normative design process has been central to automobile design for decades. While it may seem to be an unsophisticated way to design, it is important to recognize that designing a car is not the same as designing an airplane. The scale of production, the cost of the product, and the manufacturing technologies demand a completely different approach to the problem, particularly in the absence of inexpensive, widely distributed computing power. With the availability of such tools, the automakers have begun to incorporate more analytical design approaches, but the normative approach has continued to serve automobile engineering well, in the main.

Product Engineering/Manufacturing Engineering—The advance engineering group subdivides the automobile into functional subassemblies, which are passed to individual product engineering groups for final, detailed designs. The broad outlines of the advance engineering drawings are filled in, and the details of tolerance and material are worked out in the product engineering groups.

Again, past designs play a large role in defining these designs, but a harder look at the individual elements can be taken at this level. This effort will be taken, for example, when design bogeys prove difficult to meet using the historical designs. The changes may involve changing a material specification, although they usually focus upon changing the geometry of the part.

Manufacturing engineering becomes a major part of the work done by the product engineers. Although all phases of product development are geared to maintain manufacturing feasibility, the product engineers have to work closely with manufacturing engineering, not only to assure that the components that they design can be manufactured, but also to guarantee that the assembly of the rest of the car is not compromised. For example, while the engineer designing the inner panel of an automobile door might want to reduce the thickness of the panel to reduce weight, the production engineers require that the inner door has enough openings to assure that the door mechanisms can be easily installed, thus requiring a thicker, stiffer panel.

Non-OEM Contributions to Product Development— There are two major classes of actors in the automobile product design and development process who are not directly a part of the automobile companies. The first of these is the custom design house. These houses offer freelance services which support the design studios or the engineering designers. In the former case, these groups are called upon to bring particular knowledge or awareness of the automobile marketplace to enhance the appearance of the studio product. Although these operations can exist almost anywhere, they have historically been located either in Michigan, near the OEMs, or, more recently, in California, near the largest market.

Engineering support has become an increasingly important supplement to the OEM product development cycle. This is a consequence of the increasing engineering difficulty associated with the increasing demands being placed on the performance of the automobile, and the decline in the amount and breadth of engineering talent within the OEMs who have been forced to trim engineering and development effort to maintain financial goals.

These specialty engineering shops are not the only manifestation of this development. The other major actors in this area are the material and parts supplier industries. All of the major material supplier companies have followed the lead of the plastics suppliers and have made engineering, manufacturing, and design support of their material an integral part of their material selling efforts. Today, most of the major material suppliers not only offer their materials, but also finished designs of components which use these materials, backed up with complete engineering analyses. Similarly, major component and subassembly suppliers have also taken on many elements of product development and design that have traditionally been associated with the automobile companies.

In conjunction with this change has come the trend toward what has come to be called ''modular design.'' Essentially, modular design focuses on the idea that the automobile is composed of components of definable functionality which can be designed and developed in isolation from the rest of the vehicle. Although this strategy has really only come to full expression in the electrical system and in parts of the powertrain, the idea has particular attraction in the current design framework. By adopting such a strategy, a number of subassemblies or modules can be easily mixed or matched, thus retaining economies of production while offering a diverse family of products.

Summary

The process of automobile design and development is a complex endeavor that takes a product concept through stages of increasingly detailed engineering and manufacturing specifications, based on product performance and cost goals. But these performance and cost targets are affected by a number of external constraints. Because of federally mandated fuel economy and emissions requirements, environmental considerations are a major factor underlying almost every stage in the vehicle development process. This inevitably results in design tradeoffs among such factors as performance, fuel efficiency, and recyclability.

Chapter 4

Strategies for Green Design

Contents

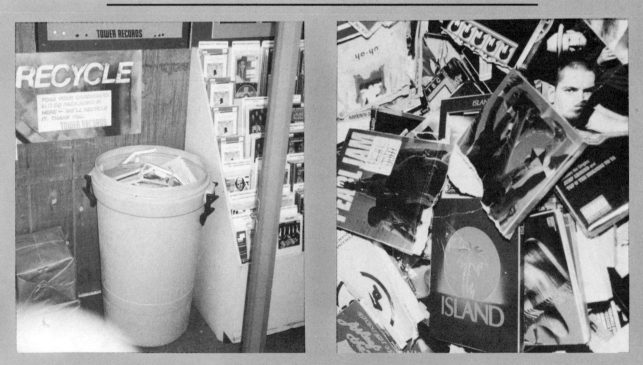

Photo credit: Tower Records

The "long box" package design for compact discs was developed to fit into existing LP record bins. The design has been criticized for its excess packaging. This example illustrates how product distribution systems can constrain design solutions. While some efforts are being undertaken to develop alternative designs that use less packaging, considerable industry resistance to a new package approach remains. Consequently, some retailers are initiating recycling programs to address consumer concerns.

Strategies for Green Design

Designers do not in general have free rein in conceiving and developing products.[1] Some constraints relate to the products themselves: for example, marketing requirements, producer capabilities, and government regulations. Other constraints relate to the systems in which products must function. Numerous examples can be cited: compact disc package specifications are determined by the size of old record bins; software applications have to conform to operating system restrictions; movie cassettes are made in VHS format despite the apparent superiority of "Beta" technology.[2] It is therefore simplistic to encourage designers to "do the right thing" without considering the constraints they face.

In fact, design choices affect—and are affected by—extremely complicated production and consumption networks (see box 4-A). As might be inferred, these networks impose constraints on the designer that have important implications when attempting to integrate environmental objectives into the design process.

Accordingly, designers can use two different strategies to integrate environmental concerns into their choices of materials and processes. One, a product-oriented approach, is to optimize the environmental attributes of the product within the constraints of the existing production/consumption network. The second strategy, a systems-oriented approach, is to broaden the scope of optimization to include changes in the production/consumption network itself. The first option is easiest, since it can be accomplished within the context of an individual firm. The second is more ambitious, because it implies a new way of looking at products, and may require new patterns of industrial organization, such as the formation of cooperative relationships among suppliers, manufacturers, and waste management providers.

PRODUCT-ORIENTED GREEN DESIGN

In the product-oriented approach, designers begin with a product concept and develop a design solution within the framework of the existing production/consumption system. Designers might ask questions such as:

- What are the waste streams from alternative manufacturing processes?
- What substitutes for toxic constituents are available?
- How is the product managed after its disposal?
- How does the design affect recyclability?
- What are the environmental impacts of the component materials?
- How is the product *actually* used by consumers?

Answering these questions may involve significant extra effort on the part of designers, and may even require new company practices, such as changing cost accounting systems to explicitly reflect a product's environmental costs, or initiating waste stream audits. But many companies are accepting this challenge, and there appear to be significant near-term benefits that could result from widespread adoption of such a design approach. For instance, at a recent conference of packaging designers, there was a consensus that—with the commitment of upper management—companies could reduce the volume of their packaging by at least 10 percent in a single year through better design.[3] Since packaging typically accounts for around 30 percent of municipal solid waste (MSW) by volume, this would mean an overall reduction of 3 percent of MSW volume in 1 year from this source alone (and perhaps a significant reduction in the industrial waste stream as well).[4]

---

[1] See the following Office of Technology Assessment contractor reports on the packaging, automobile, and electronics industries: Franklin Associates, "Packaging Design and the Environment," April 1991; Frank Field, "Automobile Design and the Environment," May 1991; Chemcycle Corp., "Environmentally Sound Product Development in the Consumer Electronics and Household Battery Industries," July 1991.

[2] Recent work provides intriguing evidence that once a particular technology path is chosen, the choice may become "locked-in," regardless of the advantages of the alternatives. See W. Brian Arthur, "Positive Feedbacks in the Economy," *Scientific American*, February 1990, pp. 92-99.

[3] Robert Hunt, Franklin Associates, personal communication, February 1991.

[4] In reality, these design changes would not all occur in a single year. Redesign can cost up to $1 million per package. Thus, manufacturers are likely to be most receptive to making changes in new package designs, or during the normal redesign cycle for existing packages.

Box 4-A—Networks

Technological advances in information technologies (computers, communications, etc.) are changing the nature of the U.S. economy, making it more complex and interdependent. These advances have led to the creation of elaborate networks that link consumers with retailers, retailers with manufacturers, and manufacturers with suppliers.[1] Virtually all sectors of the economy now depend on these production/consumption networks, with many of the networks being global in nature.[2]

As an example, consider the likely chain of events involved in the production and delivery of a frozen pizza. The pizza contains tomatoes from Mexico or California, and cheese from Wisconsin. Wheat for the pizza crust is grown in Kansas using sophisticated seeds and pesticides that are themselves the products of elaborate production networks. The pizza is assembled automatically with equipment from Germany, and wrapped in multilayered materials that are the result of considerable research and development. The pizza is probably purchased at a grocery store where a clerk passes it over a laser scanner (a device with components from Japan), which enters data into a computer and communication system designed to adjust inventories, restock shelves, and reorder products. This computer tracking system in turn makes it possible to operate an efficiently dispatched transportation system that places a premium on timely and safe delivery rather than on low bulk hauling charges. The checkout data is also used to analyze consumer response to the previous day's advertisements and to ensure that the store is closely following trends in local tastes. Any significant change in consumer buying patterns will quickly ripple throughout the production chain. So even in the case of a relatively simple product such as a pizza, the strong interconnections between disparate sectors of the economy become apparent.

Such tight linkages among industries present both opportunities and challenges for designers. Because of the flexibility they provide, networks can increase the range of possible product design options. For example, designers can choose from a wider base of materials or components suppliers. On the other hand, networks can create additional design constraints because of special distribution requirements, or greater variation in customer preferences. In many cases, networks can play a decisive role in shaping design solutions. Thus, to make significant environmental improvements, designers need to look beyond products and consider how networks themselves can be changed.

---

[1] See U.S. Congress, Office of Technology Assessment, *Technology and the American Economic Transition: Choices for the Future*, OTA-TET-283 (Washington, DC: U.S. Government Printing Office, May 1988).

[2] Organization for Economic Cooperation and Development, "The International Sourcing of Intermediate Inputs," Paris, January 1992.

SYSTEM-ORIENTED GREEN DESIGN

From an environmental perspective, it is simplistic to view products in isolation from the production and consumption systems in which they function. Is a fuel injector, for instance, a green product? From the vantage point of its component materials, probably not. But since it is designed to improve automobile fuel efficiency it could be considered "green" from a broader "systems" perspective.

Similarly, a computer, considered on its own, would probably not qualify as an environmentally sensitive product. The manufacture of a computer requires large volumes of hazardous chemicals and solvents, and heavy metals used in solder, wiring, and display screens are a significant contributor to

the heavy metal content in landfills. But the same computer could be used to increase the efficiency of a manufacturing process, thus avoiding the use of many tons of raw materials and the generation of many tons of wastes. From this perspective, the computer is an enabling technology that reduces the environmental impact of the production system as a whole.

This illustrates an important Office of Technology Assessment (OTA) finding: **green design is likely to have its largest impact in the context of changing the overall systems in which products are manufactured, used, and disposed, rather than in changing the composition of products per se.** For instance, designing lighter fast-food packaging is well and good; but 80 percent of the waste from a typical fast-food restaurant is generated

Photo credit: McDonald's Corp.

In cooperation with its suppliers, and with the assistance of the Environmental Defense Fund, McDonald's Corp. has implemented both waste reduction (left) and recycling (right) programs. By changing its sandwich packaging, McDonald's has reduced its "behind-the-counter" waste (e.g., smaller corrugated boxes) as well as its post-consumer waste.

behind the counter, where consumers never see it.[5] Addressing this larger problem requires that designers establish cooperative relationships with their suppliers and waste management providers to manage materials flows in an environmentally sound way.

Product design that accounts for the dynamic relationships among *all* companies involved in a production system has the potential to produce less waste than product design that only takes account of an individual company's waste stream. The study of the relationships among firms in production networks, and the effects of these relationships on the flow of energy and materials through our society, is an emerging field called "industrial ecology" or "industrial metabolism."[6]

The opportunities for linking product design with system-oriented thinking have not been fully explored, but examples are beginning to appear in different sectors of the economy. For instance, pesticide use has declined dramatically where farmers have adopted integrated pest management schemes involving crop rotation, and the use of natural predators.[7] Due to the success of these new methods, chemical companies are no longer simply supplying pesticides to farmers, but are also providing expertise on how to use those chemicals in conjunction with better field design and crop management. Similarly, in the energy supply sector, utilities are providing energy audit services, and are promoting customer use of energy-efficient equipment, instead of building new generating plants (see box 4-B).

A systems approach to design thus involves a unified consideration of production and consumption activities: supply-side and demand-side requirements are treated in an integrated way. This is a more far-reaching design approach in which designers might ask:

- How would new supplier and customer relationships affect the management of product materials throughout their life cycle?
- How could the same consumer need be fulfilled in a "greener" way (i.e., thinking about a product in terms of the service it provides, rather than as a physical object)?
- How could other companies' waste streams be used as process inputs?

---

[5] For example, about 35 percent of the waste generated by McDonald's restaurants is corrugated boxes, and another 35 percent is food scrap. To address these problems, McDonald's, in cooperation with the Environmental Defense Fund, has been examining the dynamics of its food distribution and production systems. By working with its suppliers to change delivery methods, and by developing composting strategies, McDonald's is taking steps to reduce these large "behind-the-counter" wastes. See the "Final Report of the Environmental Defense Fund/McDonald's Corporation Waste Reduction Task Force," Washington, DC, April 1991.

[6] Proponents of industrial ecology envision systems of production that would emulate the web of interconnections found in the natural world. See "Colloquium on Industrial Ecology," *Proceedings of the National Academy of Sciences,* vol. 89, No. 3, Feb. 1, 1992, pp. 793-884; and Robert Ayres, "Industrial Metabolism," *Technology and Environment* (Washington, DC: National Academy Press, 1989), pp. 23-49.

[7] See U.S. Congress, Office of Technology Assessment, *Beneath the Bottom Line: Agricultural Approaches To Reduce Agrichemical Contamination of Groundwater,* OTA-F-418 (Washington, DC: U.S. Government Printing Office, November 1990), pp. 115-118.

Photo credit: GE Plastics

A refillable bottling system can offer significant energy and materials savings in comparison with nonreturnable beverage containers. The impact-resistant, lightweight polycarbonate bottles shown here can be reused up to 50 times. Institutional users of refillable bottle systems, such as schools, have in some cases reduced solid waste volume by 50 percent.

- How might product design changes alter the waste stream so that it could become a useful input into another industrial process (i.e., wastes should be regarded as potential products, not just residuals of a particular industrial process)?

If the potential environmental benefits of a system-oriented approach are greater, then so are the challenges. The creation of new networks of production or distribution may be required, and long-standing relationships among manufacturers and suppliers may have to change. Such changes are not generally within the purview of product designers, and millions of dollars may be invested in the existing infrastructure for production and distribution. A systems design approach implies an unprecedented elevation of product design to the level of strategic business planning, and a new way of thinking about the environment at the highest echelons of a corporation.

Incentives for System-Oriented Green Design

There may appear to be few incentives for industry to consider such dramatic changes in existing production networks.[8] But changes of comparable magnitude are already underway. Many manufacturers are rethinking their business relationships with suppliers and customers in order to implement total quality management and concurrent engineering programs.[9] The traditional adversarial relationship between manufacturers and suppliers is giving way to a more cooperative business paradigm.[10]

General Motors, for example, has adopted an approach where it relies on a single supplier for its chemical requirements. A single chemical firm, rather than a group of suppliers, is chosen to provide, coordinate, and manage all the chemical needs of a plant and to provide continuous, on-site technical support. The supplier is remunerated according to the productive output of the plant. The supplier's profits are thereby based on the services it provides to meet a factory's production requirements, rather than the amount of chemicals sold. This cooperative strategy has reduced chemical usage by approximately 25 percent within GM facilities.[11]

The formation of environmental networks among producers, suppliers, and waste management providers could allow industry to more effectively address environmental problems. Integrated networks, in

---

[8] In fact, in some cases, there may exist regulatory disincentives. For example, it is the view of many in industry that the Resource Conservation and Recovery Act (RCRA) has impeded the reuse of spent materials. When a hazardous material falls out of a given manufacturing process, it becomes by legal definition a "waste," and is subject to stiff regulation. Because of potentially significant liability penalties, the effect of this regulation is to limit any further industrial uses of the material, and by default, the material really does become a waste. See Institute of Scrap Recycling Industries (ISRI), testimony of Herschel Cutler before the Subcommittee on Environmental Protection of the Senate Committee on Environment and Public Works, June 1991. Also see Braden Allenby, "The Design for Environment Information System," an interim report prepared for the Rutgers University Environmental Science Department, 1991.

[9] See "A Smarter Way To Manufacture," *Business Week,* Apr. 30, 1990; Genichi Taguchi and Don Clausing, "Robust Quality," *Harvard Business Review,* January-February 1990; Daniel Whitney, "Manufacturing by Design," *Harvard Business Review,* July-August 1988; "Concurrent Engineering," *IEEE Spectrum,* July 1991, p. 22; "Manufacturing: The New Case for Vertical Integration," *Harvard Business Review,* March-April 1988; "Stress on Quality Lifts Xerox's Market Share," *New York Times,* Nov. 9, 1989, p. D1.

[10] See U.S. Congress, Office of Technology Assessment, *Making Things Better: Competing in Manufacturing,* OTA-ITE-443 (Washington, DC: U.S. Government Printing Office, February 1990), p. 129.

[11] John Ogden, General Motors, personal communication, July 3, 1991.

Box 4-B—Designing a Green Energy System: Demand-Side Management

Faced with rising demand, spiraling construction costs, and strict pollution control laws, some electric utilities are trying a new strategy: they are convincing customers to buy less electricity. The strategy may seem strange, but it employs some of the central ideas of green design. In the end, it helps customers, the environment, and, surprisingly, the utilities themselves.

The strategy is known as demand-side management (DSM)—a set of techniques intended to alter how a utility's customers use electricity. Utilities using DSM do more than meet the electricity demands of their customers, they also help customers reduce or better distribute that demand. Examples of DSM include low-interest loans or rebates to homeowners who install energy-efficient heat pumps and compact fluorescent lights, and free energy audits. Like many innovative efforts at green design, DSM focuses on services, rather then goods; it encourages utilities to focus on the services provided by electricity (e.g., heating and lighting), rather than on electricity itself.

While environmentally desirable, DSM seems an unlikely strategy for a utility to pursue. Electric utilities operate as regulated monopolies, and their profits traditionally depend on sales. However, regulators in more than 30 States have adopted provisions to financially reward utilities for DSM activities. Many of these incentives treat DSM as an investment rather than an expense, allowing utilities to earn returns in the same way they do from powerplants. These regulatory incentives, coupled with high construction costs and strict pollution control laws, make DSM an attractive alternative to building new generating plants.

To influence electricity demand, many DSM programs encourage the use of energy-saving technologies. According to studies from the Electric Power Research Institute (EPRI) and the Rocky Mountain Institute, widespread adoption of these technologies could reduce total electricity demand by 24 to 75 percent. However, utility customers are often slow to adopt energy-efficient technologies, even though they provide long-term financial benefits. Customers may face institutional or financial barriers, or they may lack information on potential savings. DSM programs aim to provide information and incentives to overcome these barriers.

While DSM itself is directly applicable only to regulated utilities, its success demonstrates how economic incentives can affect large, often conservative, organizations. Some utilities undertook DSM because of its positive public relations value, but many others responded to provisions that allow them to profit directly from DSM programs. The success or failure of DSM programs may point the way toward government programs that can influence individuals and companies to adopt ''green'' technologies.

SOURCES: U.S. General Accounting Office, "Electricity Supply: Utility Demand-Side Management Programs Can Reduce Electricity Use," October 1991. Paul Klebnikov, "Demand-Side Economics," *Forbes*, Apr. 3, 1989, pp. 148-150. Leslie Lamarre, "Shaping DSM as a Resource," *EPRI Journal*, October/November 1991, pp. 4-15.

essence, expand the scale of a firm's operations and permit a firm to consider design solutions that would otherwise not be possible. In the housing industry, for example, an alliance of companies, the Integrated Building and Construction Companies (IBACoS) consortium, is developing new home concepts that promote energy and materials efficiency (see box 3-B). Ultimately, tighter inter-industry linkages could encourage the creation of closed-loop industrial systems where manufacturing byproducts from one industry are used as inputs for other industrial processes.

In Kalundborg, Denmark, an "industrial ecosystem" has been created where manufacturing wastes, surplus energy, and water are traded among a variety of different economic actors. This cooperative arrangement in energy and materials management involves a powerplant, a plasterboard maker, a cement factory, a pharmaceutical firm, an oil refinery, a collection of farmers, and the local heating utility.[12] Similar, but less elaborate efforts have been undertaken in the United States. In past years, Meridian National, an Ohio steelmaker, has sold its waste ferrous sulfate to magnetic tape manufacturers. Also, the Atlantic Richfield oil company has sold its spent silica catalysts to cementmakers. If they had not been sold, these materials would have been disposed as hazardous wastes.[13]

Product Take-Back and the Rent Model

New government regulations giving manufacturers responsibility for the environmental fate of their products are also likely to bring about systems-based

---

[12] See "A Rebirth of the Pioneering Spirit," *Financial Times*, Nov. 11, 1990, p. 15.

[13] See Robert Frosch and Nicholas Gallopoulos, "Strategies for Manufacturing," *Scientific American*, September 1989, pp. 144-152.

design solutions. For example, Germany's proposed law requiring automakers to take back and recycle automobiles has stimulated the German automobile industry to develop new cooperative strategies for auto design, manufacturing, and recycling (see box 4-C).

Perhaps the ultimate extension of the manufacturer take-back concept is the "rent model," in which manufacturers retain ownership of products and simply rent them to customers. This gives manufacturers incentives to design products to maximize product utilization, rather than simply sales.[14]

This idea was implemented in the telephone industry for many years. Before divestiture, AT&T leased virtually all telephones and thus was able to readily collect them. AT&T designed its phones with a 30-year design lifetime, and collected almost every broken or used telephone. The phones were either refurbished or were processed for materials recovery. However, with the end of AT&T's regulated monopoly and the creation of a competitive market, the number of telephone manufacturers dramatically increased. Consumers were given a wide variety of product choices. The number of phones purchased by consumers, as opposed to leased from the Bell System, grew rapidly. Accordingly, the proportion of telephones that were thrown away rather than fed back to the Bell System also increased, with a corresponding drop in the number of units available for reuse or recycling. It is estimated that approximately 20 to 25 million phones are now disposed of each year.[15]

This concept of selling *product utilization* rather than *products per se* currently applies to a variety of durable goods. Computers, copiers, aircraft, and sophisticated medical equipment are being leased rather than sold to customers. For example, Xerox leases copiers on a "total satisfaction guarantee" basis, where customers pay a certain fee for each copy and do not have to take responsibility for product operation. Some of the latest machines are even equipped with communication lines to service centers to allow automatic equipment monitoring. By retaining ownership of the products they lease, companies have a strong incentive to design goods so that they can be reused or remanufactured. In some firms it has caused a fundamental reassessment of design procedure.

When a product is viewed as an agency for providing a service or fulfilling a specific need, the profit incentive changes; income is generated by optimizing the utilization of goods rather than the production of goods.[16] While the fundamental goal of a firm would still be profit maximization, this objective could be met by marketing services as well as products. As an illustration, when a large Swiss chemical company began selling guarantees of "pest-free" fields instead of pesticides, it was able to maintain previous profit levels while reducing pesticide usage by 70 percent.[17] Thus, instead of selling as much pesticide as possible to customers, it sold a systems solution. In essence, services were substituted for chemicals.

The notion of thinking about a product in terms of the function it performs is a logical extension of total quality management (TQM) philosophy. The aim of total quality management is to satisfy customer needs. Customers usually do not care how their needs are met, as long as they are indeed met. Thus, it should not matter whether a customer's requirements are satisfied by a specific product, or by a service performed in lieu of that product.

Although the renting versus selling idea offers the possibility of reducing resource consumption rates while still meeting the needs of consumers, its range of applicability may be limited. It may work better on the corporate level than on the level of individual consumers. Average consumers may be reluctant to purchase used or refurbished goods, and divorcing products from consumer ownership could result in more careless use of those products. This model is probably more appropriate for high-value, durable products than for nondurable or disposable products.

[14] Walter Stahel, The Product-Life Institute, Geneva, personal communication, Nov. 8, 1991. For more on this idea, see Orio Giarini and Walter Stahel, *The Limits to Certainty: Facing Risks in the New Service Economy* (Boston, MA: Kluwer Academic Publishers, 1989).

[15] Braden Allenby, Senior Attorney, AT&T, personal communication, Sept. 13, 1991.

[16] Stahel, op. cit., footnote 14.

[17] Ibid.

Box 4-C—Design and Materials Management in the Auto Industry

When an old car is junked, it is often first sent to a dismantler, who removes any parts that can be resold, as well as the battery, tires, gas tank, and operating fluids. The hulk is then crushed and sent to a shredder, which tears it into fist-sized chunks that are subsequently separated to recover the ferrous and nonferrous metals.

Presently, about 75 percent by weight of materials in old automobiles (including most of the metals) are recovered and recycled. The remaining 25 percent of the shredder output, consisting of one-third plastics (typically around 220 pounds of 20 different types), one-third rubber and other elastomers, and one-third glass, fibers, and fluids, is generally landfilled. In the United States, this shredder "fluff" amounts to about 1 percent of total municipal solid waste. Sometimes, the fluff is contaminated with heavy metals and oils, or other hazardous materials.

As automakers continue to search for ways to improve fuel efficiency and reduce manufacturing costs, the plastic content of cars is expected to increase. This will not only increase the amount of shredder fluff sent to landfills, it threatens the profitability of shredder facilities, which currently depend on metals recovery to make money.

In Germany, the landfilling of old automobile hulks and the shredder residues from automobile recycling operations is a growing problem. The German Government has proposed legislation that would require automakers to take back and recycle old automobiles at the end of their lifetime. This has stimulated German automakers to explore fundamental changes in automobile design that could result in more efficient materials management. These changes would involve new relationships among auto manufacturers, dismantlers, and materials suppliers.

To avoid dealing with the auto hulks themselves, the automakers propose to take better advantage of the existing infrastructure for auto recycling. Manufacturers will design cars that can be more cheaply disassembled, and will educate dismantlers as to how to efficiently remove plastic parts. They will encourage their material suppliers to accept recovered materials from dismantlers, and will specify the use of recovered materials in new car parts, thus "closing the loop."

Green automobile design within this new framework of coordinated materials management has a very different character than auto design within an isolated firm. Instead of just thinking about how to design a fender or bumper using 10 percent less material, the designer also thinks about how the fender or bumper can be constructed from materials that can be co-recycled, and readily separated from the car body.

Several German companies, including BMW and Volkswagen, have begun to explore this system-oriented approach. BMW recently built a pilot plant to study disassembly and recycling of recovered materials, and Volkswagen AG has constructed a similar facility. The goal of the BMW facility is to learn to make an automobile out of 100 percent reusable/recyclable parts by the year 2000. In 1991, BMW introduced a two-seat roadster model whose plastic body panels are designed for disassembly, and labeled as to resin type so they may be collected for recycling.

Interest in improving materials management in the auto industry is not limited to Europe. Japan's Nissan Motor Co. has announced research programs to explore design for disassembly, to reduce the number of different plastics used, to label those plastics to facilitate recycling, and to use more recovered materials in new cars. In the United States, Ford, Chrysler, and General Motors plan to label plastic components to identify the polymers, and have recently established a consortium with suppliers and recyclers (called the Vehicle Recycling Partnership) to address the recycling issue.

Autos are already one of the most highly recycled products in the United States. This success is largely due to the efficiency of shredder technology; a single facility can process up to 1,500 hulks per day. This level of productivity is not consistent with labor-intensive disassembly operations. Although research on recycling automotive plastics is ongoing, it is not yet economically feasible to separate and recycle these materials, even when avoided landfill tipping fees are included. Thus, it seems clear that a change in materials management in the U.S. auto industry is unlikely to emerge without substantial new economic or regulatory incentives.

SOURCE: Office of Technology Assessment, 1992.

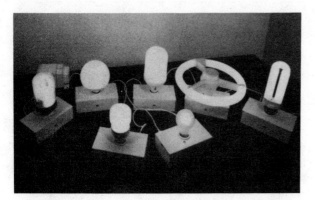

Photo credit: U.S. Department of Energy

Compact fluorescent bulbs, which are available in a variety of designs, use 75 percent less energy than standard incandescent lamps, but contain small amounts of mercury. The mercury produces the ultraviolet radiation that causes fluorescence.

MEASUREMENT ISSUES

Measuring Product-Oriented Green Design

With all of the choices available, how can designers and consumers determine what a "green" product is? As discussed in chapter 2, there may be design tradeoffs among alternative environmental attributes of a product—for instance, between waste prevention and recyclability. As an illustration, 3 pounds of a multilayered "polyester brick" packaging material can deliver the same amount of coffee as 20 pounds of metal.[18] Unlike the metal can, the polyester brick is not currently recyclable. However, to achieve equivalent levels of waste, a recycling rate of 85 percent for the metal can would be required, far higher than current rates.

Tradeoffs may also exist between other environmental attributes, such as toxicity and energy efficiency. For example, the new high temperature superconductors, which potentially offer vast im-provements in power transmission efficiency, are quite toxic; the best of them is based on thallium, a highly toxic heavy metal. Similarly, the use of compact fluorescent bulbs in lieu of incandescent bulbs can result in substantial energy savings.[19] But compact fluorescent lamps contain small amounts of mercury.[20] In this case the use of a toxic substance has measurable environmental benefits.

Life-Cycle Analysis

The existence of these tradeoffs highlights the need for analytical tools for weighing the environmental costs and benefits of alternative design choices early in the design process. One methodology that is receiving increasing attention is product life-cycle analysis (LCA). LCAs attempt to measure the "cradle-to-grave" impact of a product on the ecosystem.[21] In principle, LCAs could be used *in the design process* to determine which of several designs may leave a smaller "footprint" on the environment, or *after the fact* to identify environmentally preferred products in government procurement or eco-labeling programs.[22]

Conceptually, the life-cycle approach has helped to illuminate the environmental impacts of some products that had not been considered before, especially the "upstream" impacts associated with material extraction, processing, and manufacturing. By comprehensively accounting for materials inputs and outputs, LCAs can keep track of impacts that are merely shifted from one stage of the life cycle to another, or from one environmental medium to another. Qualitative LCAs are already being used by some companies as an internal design tool to help identify the environmental tradeoffs associated with design decisions. The life-cycle perspective also seems essential for a credible eco-labeling scheme. The first step is to develop an inventory of the

---

[18] The brick consists of polyester, aluminum foil, nylon, and low-density polyethylene laminated together. It should be noted that the coffee brick was developed to preserve product freshness, and not because of environmental considerations. See Franklin Associates, op. cit., footnote 1.

[19] Over a 10,000-hour period, *one* 18-watt fluorescent lamp replacing a 75-watt incandescent lamp results in energy savings of 570 kilowatt hours. This translates into approximately 500 fewer pounds of coal consumed, and 1,600 pounds less carbon dioxide released. Paul Walitsky, Manager Environmental Affairs, Philips Lighting Co., personal communication, May 1991.

[20] A compact bulb typically contains about 5 mg of mercury. The mercury, when vaporized in the lamp's electric arc, produces the ultraviolet radiation that causes fluorescence. (There are some data that indicate that the amount of mercury released from coal combustion for electricity generation exceeds the amount of mercury that would be used if incandescent bulbs were systematically replaced by compact fluorescents). Ibid.

[21] Ideally, the analysis consists of three steps: an inventory of resource inputs and waste outputs for each stage, an assessment of the risks associated with these inputs and outputs, and an assessment of possible options for improvement. However, virtually all LCAs attempted to date have consisted only of the first step. See Society of Environmental Toxicology and Chemistry, "A Technical Framework for Life-Cycle Assessments," Washington, DC, January 1991.

[22] "Eco-labels" are environmental seals of approval that are awarded to products whose manufacture, use, and disposal have fewer impacts on the environment than comparable products.

resource inputs and waste outputs over the entire life cycle. One approach to collecting and displaying this information is shown in figure 4-1.

If an accurate inventory can be assembled, this can provide preliminary insights into the environmental attributes of a product. But to determine definitively whether one product is "greener" than another, it is also necessary to know how the quantities in figure 4-1 should be weighted to reflect their relative health and environmental risks. For example, how should a pound of sulfur dioxide emitted to the air during manufacture be compared with a pound of solid waste going to the landfill? Is it more desirable to use laundry detergents that contain phosphates or phosphate-free detergents that release volatile organic compounds? To resolve such questions, additional information about environmental fate, exposure pathways, and dose-response data for each environmental release is required.

Another serious limitation is that the data requirements of a comprehensive LCA can quickly get out of hand. A major problem is where to draw the boundaries of the analysis. Can certain materials and energy flows be ignored without overlooking some significant environmental effects? For example, should one consider the energy required to produce the fertilizer that is used to grow the cotton that is used in cloth diapers? And if the energy is derived from coal as opposed to hydroelectric power, should one count the sulfur dioxide emissions associated with the combustion of coal?[23] Moreover, data uncertainty can be compounded by the fact that life-cycle analysis is sensitive to changes in inputs over time. If a few material or technology inputs change, initial assumptions may no longer hold, and the inventory might require a complete updating. When applied to more complicated products like televisions and automobiles, the LCA methodology might become hopelessly difficult to implement. Precisely because LCAs are multidimensional, interest groups are free to emphasize the aspects most favorable to their own agendas, thus providing almost limitless potential to confuse consumers.[24]

Before LCAs (inventory and risk assessment) can become a complete tool for comparing the greenness

Figure 4-1—Life-Cycle Inventory

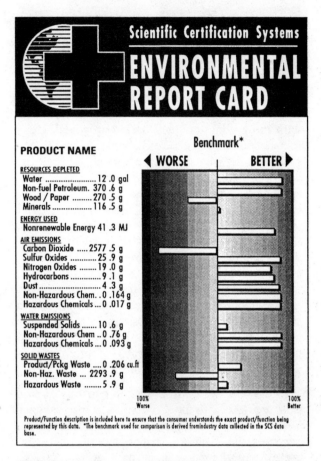

One approach to reporting the results of a life-cycle inventory is illustrated by this hypothetical comparison of a product against a benchmark product. For each product, the resource depletion, energy consumption, air emissions, water emissions, and solid wastes associated with manufacture and product use are tabulated. The inventory approach shown here will be used by SCS in lieu of a simple eco-label.

SOURCE: Scientific Certification Systems.

of products after the fact, these issues will have to be resolved. Less information will probably be required, not more. LCAs may have to be streamlined to focus on a few critical dimensions of a product's environmental impact, rather than all dimensions. One possibility might be to limit the analysis to three dimensions: a product's contribution to catastrophic or irreversible environmental impacts (e.g., ozone destruction, species extinction), acute hazards to human health, and life-cycle energy consumption.

---

[23] These methodological issues will be discussed in the upcoming Environmental Protection Agency report, "Life-Cycle Assessment: Inventory Guidelines and Principles," Battelle and Franklin Associates, contractor report for U.S. EPA Office of Research and Development.

[24] "Life-Cycle Analysis Measures Greenness, But Results May Not Be Black and White," *Wall Street Journal*, Feb. 22, 1991, p. B1.

Any such "partial" LCA can be criticized as being incomplete; for example, according to the criteria above, chronic health effects of long-term exposure to low concentrations of chemicals would not be considered. But some such simplification seems essential if LCAs are to be widely used.

There are further difficulties. Because they are inherently product-focused, LCAs are consistent with the product-oriented design approach. But in focusing attention on the environmental attributes of products *per se*, the LCA approach to design may divert attention from larger opportunities available by designing products in concert with new systems of production, consumption, and waste management.[25] Fundamentally, LCAs are "static" in that they provide a snapshot of material and energy inputs and outputs in a dynamic production system. LCAs therefore do not capture the opportunities for new technologies and new production networks to reduce resource use and wastes. In assuming that the product will be offered with certain characteristics to perform a certain service, LCAs may limit the scope of designers to consider ways of providing the service in more environmentally sound ways.

In the near term, life-cycle comparisons of products are likely to be limited to comparisons of resource and waste inventories. For designers' purposes, the inventory need not be exhaustive to be useful. For the purpose of product labeling, the inventory should be rigorous, easily verifiable, and periodically updated. Even so, at best, the inventory will clarify environmental tradeoffs, rather than provide definitive conclusions.

Measuring System-Oriented Green Design

How do we measure the environmental impact of alternative systems, as opposed to alternative products? Product characteristics are tangible and can—at least in principle—be quantified through life-cycle analysis; systems characteristics are less tangible. To measure the environmental performance of systems (say, transportation systems or telecommunication systems), new metrics will be required. Perhaps aggregate indicators like "energy intensity" and "materials intensity" could be used to compare "green systems."[26]

Another useful aggregate measurement tool might be provided by input-output analysis. Input-output analysis models the exchanges (inputs and outputs) between producers and suppliers. It can be used to examine the exchanges among a small group of companies, or the workings of national economies. In principle, input-output techniques could be used to correlate both intermediate and final products with emissions of various pollutants.[27] Using these models, it might be possible to track the pollution associated with alternative production networks.

With an emphasis on service, we may be more concerned about product utilization rates rather than disposal rates or quantities of emissions. For instance, a measure of environmental performance might be product lifetime, or how effectively a product performs its designated task (e.g., the efficacy of a pesticide as part of an integrated pest management scheme). Credible measurement tools to evaluate the environmental performance of networks are an important research topic for the future, as discussed in chapter 6.

SUMMARY

Green design thinking can occur on several levels. At the product level, designers can optimize designs so as to improve materials and energy efficiency, or product longevity. A more ambitious approach is to think about how product designs might be optimized in a context of reorganized production and consumption systems. Such an approach suggests a design philosophy that places primary emphasis on the service a product provides rather than the product itself. Thus, systems solutions require real behav-

---

[25] Is it correct to consider the LCA results of a particular product in isolation from the ripple effects of that product in the economy? The environmental externalities associated with a product might be outweighed by the greater efficiencies achieved when that product is incorporated into other products; a good example would be computer chips and the automated systems that use those chips to improve manufacturing efficiency.

[26] Energy intensity refers to the Btus used to produce a dollar's worth of gross domestic product (GDP); materials intensity refers to the quantities of materials (metals, lumber, cement, etc.) used to produce a dollar's worth of economic output. In recent decades, both energy intensity and materials intensity in the United States have declined. See U.S. Congress, Office of Technology Assessment, *Energy Use and the U.S. Economy*, OTA-BP-E-57 (Washington, DC: U.S. Government Printing Office, June 1990); and Eric Larson, Marc Ross, and Robert Williams, "Beyond the Era of Materials," *Scientific American*, June 1986, pp. 34–41.

[27] With the availability of mandated databases on industrial waste streams such as EPA's Toxics Release Inventory, it is becoming feasible to incorporate pollution data into these economic models. See Faye Duchin, "Industrial Input-Output Analysis: Implications for Industrial Ecology," op. cit., footnote 6, pp. 851–855.

ioral change on the part of producers and consumers, and can be difficult to implement. However, if the systems in which products are manufactured, used, and disposed can be modified, the environmental benefits will likely go well beyond what is possible by focusing on products alone.

International Comparison of Policies Affecting Green Design

Contents

Box

Figure

Table

International Comparison of Policies Affecting Green Design

In recent years, interest in green product design has increased dramatically in the United States and other industrialized nations. Historically, environmental policies have focused on protecting air, water, and land from "point" sources of pollution (e.g., factories and powerplants). But countries are now recognizing the importance of nonpoint sources, including products (e.g., chlorofluorocarbons (CFCs) and pesticides). Many countries with stringent environmental protection standards show a growing tendency to extend the traditional emphasis on pollution control to include standards for the environmental attributes of products.

The environmental product policies of other nations have important implications for the United States. First, these policies are shaping international markets in which U.S. goods must compete. The policies of other nations on issues such as packaging, mercury in batteries, and automobile recycling have the potential to change the competitive landscape of foreign markets. The success of U.S. companies in these markets will depend, at least in part, on their ability to employ green product design.

Second, product policies may act as nontariff barriers to trade.[1] They are often seen by critics as giving domestic industries an unfair advantage. Examples include the recent U.S. attempt to ban imports of Mexican tuna because of concern about dolphins killed during tuna fishing,[2] Denmark's decision to ban the sale of beer in nonrefillable containers, and Germany's new law requiring companies to recover and recycle their packaging waste.[3] Finally, studying the experience of other industrialized nations can provide lessons for U.S. policymakers.

This chapter surveys some of the more notable policies affecting green design in industrialized countries, with a view toward understanding how

U.S. activities compare with those in other countries.[4] Table 5-1 provides a summary.

ENVIRONMENTAL POLICIES AFFECTING PRODUCT DESIGN

Europe

Nearly all European countries are building up a body of product-related environmental law that extends beyond traditional areas of pollution control. There is a strong positive correlation between national wealth and environmental awareness in Europe; thus Germany, the Netherlands, and the Nordic countries (Norway, Sweden, Finland, and Denmark) are in the forefront.

Germany

Germany is particularly active in environmental product policy. In May 1991, Germany enacted its Waste and Packaging Law that gives manufacturers and retailers responsibility for recovering and recycling their own packaging wastes (see box 5-1). This statutory coupling of manufacturing with postconsumer recycling forces manufacturers to account for the solid waste implications of packaging. Germany is considering similar laws that would give manufacturers the responsibility for collecting and recycling durable goods, such as household appliances and automobiles (see below).

Besides recycling, Germany has an active program for labeling environmentally preferred products. The "Blue Angel" eco-labeling scheme has been in operation since 1978 and is the only example of a well-established eco-labeling scheme in Europe. The award is not given to individual products, but to categories of products that meet certain criteria. Supporters of the Blue Angel scheme point to several successes: paint, lacquers, and varnishes that are low in solvents and other hazardous substances

---

[1] For an overview of the issues involving trade and the environment, see: U.S. Congress, Office of Technology Assessment, *Trade and Environment: Conflicts and Opportunities*, OTA-BP-ITE-94 (Washington, DC: U.S. Government Printing Office, May 1992).

[2] The ban was imposed under the Marine Mammal Protection Act of 1972, Public Law 92-552. Later, a panel of the General Agreement on Tariffs and Trade (GATT) determined that the ban violated GATT's rules of international trade. Ibid.

[3] Frances Cairncross, *Costing the Earth: The Challenge for Governments, the Opportunities for Business* (Boston, MA: Harvard Business School Press, 1992).

[4] The discussion of activities in foreign nations draws heavily from Environmental Resources Limited, *Environmentally Sound Product Design: Policies and Practices in Western Europe and Japan*, contractor report prepared for the Office of Technology Assessment, July 1991.

Table 5-1—Environmental Policies Relating to Products in Other Industrialized Countries

Policy	Comment
Economic Commission for Europe (United Nations) A task force is developing guidelines for "environmental product profiles," a qualitative description of the environmental impacts of a product for use by commercial and institutional buyers.	Researchers from the Netherlands and Sweden have been among the most active participants in the Task Force.
European Community Draft law requiring specific percentages of recovery (recycling, incineration, and composting) for product packaging.	The draft has been driven largely by German packaging legislation.
EC eco-label.	Principles of the program have been agreed upon, but no date for implementation has been set.
EUREKA Eco-design project to gather information and develop methods to stimulate the design of environmentally sound products.	Under the Euro-Environ umbrella program, this project is led by Dutch researchers.
Canada The National Packaging Protocol is a voluntary program with packaging reduction targets and dates.	By the year 2000, packaging sent to disposal is to be no more than 50 percent of the amount sent in 1988. Half of this reduction is targeted to come from waste prevention and re-use, and half from recycling. Regulations are to follow if targets are not achieved.
Environmental Choice eco-label.	Over 400 product categories have been recommended for labeling.
Denmark Ban on domestically produced nonrefillable bottles and aluminum cans.	The EC took Denmark to the European Court over this ban, which originally covered foreign-produced containers as well, claiming it was an unwarranted restriction on trade. Denmark won the case. Now, deposit, return, and recycling schemes must be set up for imports.
Fee imposed on waste delivered to landfills and incinerators as an incentive to recycling and to support clean technology.	130 DKr ($19) per ton is earmarked for subsidies for clean technology.
Clean Technology Action Plan (1990-92).	A principal aim of the plan is to reduce consumption of nonrenewable materials and to reduce the use of heavy metals and other toxic substances.
Germany Packaging Waste Law, passed in 1991, gives manufacturers responsibility for collecting and recycling various kinds of packaging at specified rates by certain dates.	This legislation is being considered as a model for EC-wide packaging legislation. The packaging collection rates and target dates are considered very ambitious. Concerns have been raised that this law could create special problems for imported goods.
Manufacturer take-back-and-recycle laws have been proposed by the government for automobiles, electronic goods, and other durables.	These proposals, which have not yet been passed, would go into effect in 1994. They have already stimulated auto and computer companies to begin to redesign cars and computers to facilitate recovery and recycling of components and materials.
Mandatory deposit refund on plastic beverage containers (except milk).	Established in 1989, the deposit of DM 0.5 ($0.28) will remain in place under the new packaging waste law.
Blue Angel product eco-label.	Begun in 1978, this was the first national eco-label program; it now covers 400 products in 66 categories.

Japan

Recycling Law, passed in 1991, sets target recycling rates around 60 percent for most discarded materials by the mid-1990s. Includes product redesign strategies for packaging and durable goods.

The law gives the Ministry of International Trade and Industry broad powers to set recycling guidelines for specific materials and industries.

Eco-mark product eco-label.

The label covers more than 850 products in 31 categories.

Netherlands

National Environmental Policy Plan sets national targets and timetables for implementing clean technology, including redesign of products.

The most comprehensive national environmental policy planning document anywhere in the world. The Netherlands Government has a budget of around $100 million per year to support development of clean technologies and products.

Voluntary agreements reached with industry targeting 29 priority waste streams and reduction of packaging waste.

Voluntary agreements are considered a more effective means of achieving environmental goals than command-and-control regulations.

Norway

Tax on nonreturnable beverage containers.

This tax, which can be as high as $.52 per container, is intended to encourage producers to use refillable packaging.

Deposit-refund on old car bodies.

The deposit of NKr 1,000 (U.S. $143) is refunded with a bonus; the return rate is 90 percent.

Sweden

Ban "in principle" on the use of cadmium.

A number of exemptions are permitted.

Voluntary deposit-refunds for glass and aluminum beverage containers.

Return rates of 80 to 90 percent have been achieved.

United Kingdom

Gas tax differential of around 10 percent between leaded and unleaded gas.

Several other countries have similar policies. Sales of unleaded gas rose from negligible to 36 percent in 3 years.

SOURCE: Office of Technology Assessment, 1992.

Box 5-A—Germany's Packaging Law

Germans generate about 32 million tons of municipal waste per year. About 30 percent of that waste is incinerated and nearly all the rest, about 22 million tons annually, ends up in landfills. At this rate, about half of Germany's landfills will fill to capacity and be forced to close within 5 years.

Because packaging accounts for 30 percent of German municipal waste by weight, the country recently enacted a prominent new law regarding the collection and recycling of packaging. The law (The German Federal Ordinance Concerning Avoidance of Packaging Waste) gained final legislative approval in April 1991, and its first provision took effect in December 1991. The law redefines the responsibilities of companies and requires recycling on a massive scale.

The fundamental philosophy behind the German packaging law holds product manufacturers and distributors responsible for the packaging they create and use. The law requires little from consumers, but mandates that companies take back and recycle used packaging. For some types of packaging, the law gives industry an opportunity to establish its own collection and recycling system. If such self-management fails, however, it compels manufacturers and distributors to collect the packaging themselves and arrange for recycling.

The law defines three types of packaging: transport, secondary, and sales. Transport packaging refers to items used to protect or secure products during transportation from the manufacturer to the distributor (e.g., large corrugated shipping containers and wooden pallets). Secondary packaging refers to items used to group, protect, and display the product at the point of sale (e.g., exterior cartons and packaging components that make products tamper-proof). Sales packaging refers to items in direct contact with the product itself (e.g., liquid containers and food wrapping).

The law contains separate deadlines for each type of packaging. Collection of transport packaging by manufacturers and distributors was required beginning December 1, 1991. Collection of secondary packaging by distributors was required beginning April 1, 1992. Sales packaging must be taken back beginning January 1, 1993.

Collected packaging must be reused or recycled to the greatest extent possible. Materials not recycled or reused must be materials that: 1) cannot be separated manually or by machine; 2) are soiled or contaminated by substances other than those that the package originally contained; or 3) are not integral parts of the packaging. Recycling must be accomplished independently of the public waste disposal system. Incineration is specifically prohibited.

Additional provisions apply to specific packaging types. Secondary packaging must be removed by distributors (including retailers) before products reach consumers or distributors must provide an opportunity for consumers to remove and return the packaging at the point of sale. The law requires that distributors provide separate containers for different packaging materials and post signs indicating that consumers may return secondary packaging.

The law also contains additional provisions for sales packaging. As with secondary packaging, distributors must accept returned sales packaging at the point of sale. The law also mandates a deposit-refund scheme covering containers for beverages, household cleaners, and spray paints.

The sales packaging provisions can be avoided by manufacturers and distributors who are party to an alternative collection system. First, the alternative system must collect packaging directly from households or establish collection centers. Second, the system must meet strict collection and sorting targets. These targets will be assessed by weight within each "Lander" or district within Germany, and require at least 60 percent collection of most materials by January 1993 and at least 80 percent collection of all materials by July 1995. Third, existing levels of reusable beverage containers must be maintained. This alternative system exempts companies from the provisions for sales packaging only; companies must still take back transportation packaging and secondary packaging directly.

Under pressure from retailers, industry moved rapidly to establish an alternative system under the terms of the law: the Duales System Deutschland (DSD). DSD is a private company established to collect packaging of participating companies. Participating companies pay a licensing fee to use a "Green Dot" label that identifies their packages as eligible for collection. Licensing fees of up to 20 pfennig (U.S. $0.12) per package are expected to raise about 2 billion DM (U.S. $1.2 billion) per year.

Whether reality will match the law's lofty goals remains to be seen. The law faces several hurdles. First, the mandatory 1995 collection rate of 80 percent for all sales packaging materials far exceeds the rates currently achieved. For example, Germany recycled just over 40 percent of paper and paperboard recycling in 1987 and just over 53 percent of glass in 1989. Internationally, 80 percent recycling rates for any material are rare, even in highly motivated neighborhoods. Overall rates of 80 percent are unheard of on a national scale.

Second, although the stated goal of the law is source reduction, it focuses almost exclusively on recycling. Whether the costs of collection and transportation will encourage source reduction remains to be seen. Third, the law does little to enlist the help of consumers in recycling. The entire burden for ensuring the success of the law rides on the efforts of manufacturers and retailers. Finally, the law raises thorny issues regarding international trade. The law's provisions apply to any goods sold within Germany, regardless of their country of origin. Thus, companies that export goods to Germany must arrange for collection and recycling of their packaging.

While the European Community has been working on unified solid waste guidelines to facilitate free trade, the German law has leapt ahead with the strictest plan of any EC nation. Whatever the outcome, Germany's packaging law represents a bold experiment that will be closely watched on both sides of the Atlantic.

SOURCES: James E. McCarthy, *Recycling and Reducing Packaging Waste: How the United States Compares to Other Countries*, 91-802 ENR (Washington, DC: Congressional Research Service, Nov. 8, 1991). "Translation of the Ordinance on the Avoidance of Packaging Waste" in Environmental Resources Limited, *Environmentally Sound Product Design: Policies and Practices in Western Europe and Japan*, contractor report prepared for the Office of Technology Assessment, July 1991. "Recycling in Germany: A Wall of Waste," *The Economist*, Nov. 30, 1991, p. 73. Kerstin Wessel, "The German 'Dual System'—An Instrument To Promote Waste Minimization in the Packaging Sector?" *Packaging and the Environment—Policies, Strategies and Instruments*, Invitational Expert Seminar, Trolleholm Castle, Sweden, Feb. 7-8, 1991 (Lund, Sweden: Department of Industrial Environmental Economics, Lund University).

now command 50 percent of the German do-it-yourself market, compared with just over 1 percent in the 1970s; over the same period, emission standards for oil and gas heating appliances have improved by more than 30 percent. The program receives only 8 percent of its income from Federal subsidy, with 57 percent coming from the sale of publications and certification.[5]

However, the Blue Angel program is not an unqualified success. Despite its longevity, the program only covers a small percentage of consumer products.[6] Although the initial intention was to consider all of a product's environmental impacts when awarding the Blue Angel label, in practice attention usually focuses on one or two environmental impacts. For example, the program judges spray cans on the elimination of aerosol propellants and judges detergents on wastewater load. Environmental groups have criticized the program, contending that it should consider the entire product life cycle. The feasibility of broadening the selection criteria to include life-cycle impacts is presently under study.

The Netherlands

The Dutch Government produced the National Environmental Policy Plan (NEPP) in 1989 and NEPP Plus in 1990. These are major policy documents outlining plans for harmonizing economic development with the environment through the year 2010. They are widely acknowledged to be the most detailed and comprehensive example of environmental planning anywhere in the world.[7] The plans explicitly include product policy and green design as part of a preventive strategy using "process-integrated environmental technology" to achieve "sustainable development." The Dutch plan looks forward to "an alternative way of living" with investment in "clean" technologies coming to dominate new capital investment. The Dutch budget for development of clean technology was about $90 million in 1990. No other country has long-term policies that address environmental aspects of product design as specifically as the Netherlands.

Dutch environmental policy relies increasingly on voluntary agreements negotiated with industry, rather than on command-and-control regulation. For exam-

---

[5] Ibid., pp. 11-12

[6] Of the labels issued, over half have been in only four product categories (recycled paper, low-pollutant varnishes and coatings, low-emission gas burners, and pH neutral stripping agents for wastewater treatment). *Environmental Labelling in OECD Countries* (Paris: Organisation for Economic Cooperation and Development, 1991), p. 48.

[7] ERL, op. cit., footnote 4, p. 22.

Figure 5-1—Eco-labels Around the World

Canada (Environmental Choice)

Nordic Countries (White Swan)

West Germany (Blue Angel)

Japan (EcoMark)

United States (Scientific
Certification Systems)*

United States (Green Seal)

Eco-labels are intended to identify environmentally preferred products for consumers. Above are government-sponsored labels from four foreign programs and two private U.S. labels.

*NOTE: The SCS label will provide comparative data on environmental attributes (see figure 4-1).

ple, it is part of government policy to identify hazardous substances (e.g., cadmium and chlorine),

and to eliminate these substances from every stage of the production process. The Dutch Government has established waste reduction targets for 29 priority waste streams, with action plans to be negotiated as voluntary agreements with industry. The government has recently signed a voluntary agreement on packaging waste, which could be backed up by regulations if negotiated targets are not met within the specified time. Environmental groups in the Netherlands also negotiated a voluntary agreement with retailers on the elimination of polyvinyl chloride (PVC) from packaging in September 1990.

The Netherlands has begun work on a national eco-labeling scheme for products, but has expressed a preference for a harmonized European Community (EC) program. Dutch researchers are active in the area of life-cycle analysis. The Netherlands initiated a task force under the aegis of the United Nations Economic Commission for Europe (ECE) to develop guidelines for product "profiling," a descriptive form of life-cycle analysis intended for use by professionals such as designers and procurement agents. In addition, the Dutch have initiated research projects to develop guidelines and information resources to assist designers in making better environmental choices.[8]

Nordic Countries

Sweden, Norway, and Denmark have all been active in the area of environmental product policy. For many years, these countries have employed a wide range of taxes and deposit-refund schemes to limit packaging waste (see below). They have also used a combination of bans and voluntary agreements with industry to encourage green design.[9] Sweden banned "in principle" the use of cadmium in many products over a decade ago, and the idea of "sunsetting" or phasing out the use of various toxic chemicals is quite popular there. Denmark banned nonrefillable beverage containers for beer and soft drinks, and required that bottle designs be government-approved.[10] Denmark has announced a Clean Technology Action Plan to run through 1992 that focuses on reduced consumption of nonrenewable materials

---

[8] ERL, op. cit., footnote 4, p. 23.

[9] See Christian Ege Jorgensen, "Sunset Chemicals—From a Danish Perspective," *Proceedings of the Global Pollution Prevention '91 International Conference & Exhibition*, Lorraine R. Penn (ed.), Washington, DC, Apr. 3-5, 1991.

[10] Denmark was taken to the EC Court over the ban, which allegedly constituted a restriction on trade. The court ruled in favor of Denmark, but required that designs of imported bottles be exempted from the approval requirements as long as they were nonmetal and were subject to a deposit-refund scheme.

and reduced use of heavy metals and other toxic substances in a variety of products.

The Nordic countries decided in November 1989 to "implement a harmonized, voluntary, and positive Nordic environmental labeling system for products."[11] Criteria for product categories are currently being drawn up, although this program is now at least partially on hold, pending an EC decision on an EC-wide labeling system.

European Community

The European Community is increasingly a driving force behind environmental law in Europe. The number of EC environmental laws adopted has risen from one per year in the 1960s to between 20 and 30 per year in the 1980s.

EC environmental policy has always included a strong emphasis on harmonizing product standards among countries, but such harmonization can be difficult. Much EC environmental legislation and planning has been inspired by Germany (e.g., manufacturer responsibility for packaging waste), the Netherlands (e.g., waste stream prioritization), and the Nordic countries (e.g., reduced heavy metal content and separate collection of batteries). However, the range of environmental legislation varies greatly among European countries, as does the willingness of different countries to pursue future action.

Some countries already have substantial environmental standards in place, and have expressed fears that their own higher standards may be compromised by a lower EC-wide standard.[12] As a result, the EC tends to set minimum standards that may be exceeded by "greener" countries. Some poorer countries have not placed as high a priority on environmental policy, and they find their national legislation increasingly driven by EC requirements. While EC standards have tended to be relatively stringent, they often acknowledge that poorer countries may have difficulty meeting the standards. The EC sometimes adopts a two-tier approach that gives poorer countries more time to achieve standards demanded by wealthier countries.

The EC began discussions on eco-labeling in 1988, prompted by the need to coordinate product labeling before the advent of the Single European Market in 1992. Plans to launch the system were formally unveiled in November 1990. The criteria for granting the label will be harmonized throughout the EC and will be decided by the European Commission with the assistance of an Advisory Committee. The decision to adopt life-cycle criteria was made after pressure from the Nordic countries (though they were not EC members), who expressed criticism of the more limited criteria applied by Germany's Blue Angel program. At this writing, no date had been set for implementing the EC eco-label (a daisy surrounded by 12 stars).

Japan

Japanese industry's interest in green products has lagged somewhat behind that in Europe and the United States, in part because of a lower level of consumer activism, and in part because there has been little government policy leadership in this area. However, Japan recently identified environmentally sound products and technologies as a major new market opportunity, and is investing large sums in research and development.[13] The close relationship between government and industry in Japan suggests that government proposals—as they develop—may be implemented more quickly than is the case in Europe or the United States.

One motivation for Japanese environmental policy is an acute crisis of landfill space. Although Japan incinerates 70 percent of its municipal solid waste, major urban areas are having difficulty even finding space to dispose of the incinerator ash residue.[14] In response, the government passed a recycling law in April 1991 that is designed to promote waste recycling.[15] The recycling law mandates recovery rates of around 60 percent for most discarded materials (including glass, paper, aluminum cans, steel cans, and batteries) by the mid-1990s, and it includes product redesign strategies for

---

[11] Nordic Council of Ministers, written procedure, Nov. 6, 1989. Cited in ERL, op. cit., footnote 4, p. 12.

[12] For example, the court case stemming from Denmark's ban on nonrefillable bottles.

[13] Jacob M. Schlesinger, "Thinking Green: In Japan, Environment Means an Opportunity for New Technology," *Wall Street Journal*, June 3, 1992, p. A1. Neil Gross, "The Green Giant? It May Be Japan," *Business Week*, Feb. 24, 1992, pp. 74-75.

[14] ERL, op. cit., footnote 4, p. 74.

[15] It is formally known as the "Law for Promotion of Utilization of Recyclable Resources," or more commonly as the "Recycling Law."

both packaging and durable goods. Sponsored by the Ministry of International Trade and Industry (MITI), the law provides MITI with broad powers to set recycling guidelines for specific industries and materials. Those interviewed by the Office of Technology Assessment felt that MITI's involvement, as well as its extensive discussions with industry prior to the law's passage, would mean that Japanese industry would move relatively quickly to implement the law.[16]

The ability to move quickly was certainly illustrated in the case of Japan's environmental labeling scheme, the Eco-mark. The details of the scheme were published in February 1989, along with an initial list of approved product categories. In March 1989, 46 products in 7 product categories were approved, including aerosols containing no CFCs. One year later, there were 850 labeled products in 31 categories.[17] The Eco-mark is usually awarded to product categories based on a single environmental attribute, and thus is less rigorous than criteria proposed in EC draft legislation or other national labeling schemes based on the cradle-to-grave approach. However, the Eco-mark was quickly implemented and is reportedly popular with Japanese consumers, who have not traditionally been associated with strong environmental awareness.

ANALYSIS

The integration of product policy into environmental policy and the role of product design in making products more "friendly" to the environment are areas of considerable policy ferment around the world. Twenty-two of the major industrialized countries either have a national eco-labeling program for products, or will have one soon.[18] There are a growing number of product control policies in effect, ranging from outright bans on materials to economic instruments such as product taxes (see table 5-1). All of the countries are attempting to boost recycling; many of these initiatives focus on packaging, which constitutes about one-third of post-consumer waste by weight in many countries.[19]

There are some important differences between the U.S. approach and the approach taken by other countries. In some countries, environmental and/or economic conditions have forced policies that encourage green design. While shrinking permitted landfill capacity is a growing problem in the United States, it is already very serious in Northern Europe and Japan. As a result, the pressure on manufacturers to design smaller, more efficient products and packages is greater than in the United States. Another difference is the dramatically higher fuel prices in Europe and Japan, due in large part to government taxes.[20] These high prices encourage the design of fuel-efficient automobiles, contribute to greater use of public transportation, and promote more energy-efficient buildings and appliances.

The political atmosphere surrounding waste management in Europe has forced drastic policy measures such as Germany's Packaging Waste Law. This law has set the tone for a common policy theme emerging in several European countries: the idea of giving manufacturers responsibility for recovering and recycling their products at the end of their useful life. Manufacturer take-back requirements have intuitive appeal because they give designers direct incentives to consider how the product will be recovered and recycled, thus "closing the loop" among design, manufacturing, and waste management.

The idea of shifting responsibility for managing these materials to manufacturers can be expected to have a growing appeal in the United States as well, particularly given that U.S. cities are collecting recyclable materials at a rate much faster than they are being used. Many U.S. manufacturers, especially those of durable goods, feel that similar legislation is inevitable in the United States in a few years.

There are also social and cultural differences in Europe and Japan that may foster the development

---

[16] ERL, op. cit., footnote 4, p. 76.

[17] ERL, op. cit., footnote 4, p. 80.

[18] Catherine Arnst, "Some 22 Nationals Could Have 'Green Label' Schemes by '93," *Toronto Star,* Nov. 6, 1991, p. D6.

[19] James E. McCarthy, *Recycling and Reducing Packaging Waste: How the United States Compares to Other Countries*, 91-802 ENR (Washington, DC: Congressional Research Service, Nov. 8, 1991).

[20] For example, gasoline prices in European nations are two to four times the price of gasoline in the United States, with almost all of the difference due to government taxes. Japanese gasoline prices are more than three times higher, with about half of the difference due to taxes. Energy Information Adminstration, *Indicators of Energy Efficiency: An International Comparison*, EIA Service Report, SR/EMEU/90-02, July 1990.

of green design more rapidly than in the United States. In many other countries, government and industry work together more comfortably than in the United States, where the relationship tends to be more confrontational. Although Europe continues to rely heavily on command-and-control environmental regulations, the greener European countries are more likely to seek voluntary agreements with industry to achieve environmental goals, rather than enforcing compliance with regulations through the legal system. Large government subsidies to industry for development of "clean technology" are prevalent in the Netherlands and Denmark.[21] The closer relationship between government and industry may explain why several other countries have national eco-labels for products, while the United States leaves labeling efforts to the private sector.

Attitudes toward weighing environmental risks and benefits also differ somewhat between the United States and some European countries. New initiatives in Europe, such as the German packaging waste law or carbon taxes on fuels, tend not to be subjected to the kind of cost-benefit analysis that would be expected in the United States.

Finally, in countries like Germany, Sweden, Denmark, and the Netherlands, the policy debate is qualitatively different from that in the United States. These European nations produce national policy documents that state broad environmental goals such as resource conservation and "sustainable economic development"—with explicit targets and timetables. U.S. policies focus more narrowly on protecting consumers from harmful products and protecting the environment from various waste streams. Using the terminology of chapter 3, these countries are developing policies from the perspective of the resource management and eco-development paradigms, while the United States is operating from the environmental protection paradigm.

With the approach of the Single Market in 1992, the member countries of the European Community are wrestling with the problem of harmonizing their different environmental product standards and recycling laws. These laws have proved contentious in the past, and no resolution is in sight.[22] The United States faces similar problems in managing the multitude of divergent environmental product regulations in various States. Recent controversies over whether countries can restrict imports of goods deemed harmful to health or the environment, or whether such restrictions constitute nontariff barriers to trade, suggest that the harmonization of international environmental product policies will be a thorny problem for future negotiations under the General Agreement on Tariffs and Trade (GATT) and other international agreements.

CONCLUSION

On the whole, the United States cannot be said to be "behind" other countries in the development of environmental policies that encourage green product design. Indeed, many European countries look enviously at U.S. environmental policies such as auto emissions standards, or the timetable for phaseout of CFC production and use, which are among the most aggressive in the world. Some U.S. companies are acknowledged world leaders in waste prevention techniques.

After investigating the policies of other nations, the Office of Technology Assessment finds no models that the United States should directly imitate. In fact, many observers believe that some of the more extreme measures, such as Germany's mandatory take-back provisions for packaging waste, will prove to be costly and difficult to implement.[23] This does not mean that the United States should ignore the potential of green product design, only that the policies pursued abroad should not be copied wholesale.

The rapid evolution of environmental product policy, and its increasingly international flavor, suggests that the United States needs a proactive Federal involvement. First, such involvement can ensure that the experiences of other nations are closely monitored. Second, Federal involvement can provide a focal point for policies that protect the environment while reducing barriers to international trade. In the next chapter, options for greater Federal involvement are discussed.

---

[21] For example, the Netherlands Government provided $90 million in 1990 to subsidize clean technology development. As a percentage of gross national product, this would be the equivalent of about a $2 billion program in the United States.

[22] Frances Cairncross, "How Europe's Companies Reposition To Recycle," *Harvard Business Review*, March-April 1992, p. 34.

[23] "Environmentalism Runs Riot," *The Economist*, Aug. 8, 1992, pp. 11-12.

Policy Options and the Challenge of Green Design

Contents

Policy Options and the Challenge of Green Design

Several examples have already been cited in which Federal regulations inject environmental considerations into product design. For instance, the effect of Corporate Average Fuel Economy (CAFE) standards on automobile design is described in appendix 3-A. Should Congress consider taking any further action to encourage green design? While some in industry argue that existing market incentives and environmental regulations are sufficient, the Office of Technology Assessment (OTA) finds that further Federal action is necessary to ensure that the full potential of green design is realized. This chapter examines current incentives for green design and identifies four areas of need that only Congress can address.

CURRENT INCENTIVES FOR GREEN DESIGN

Federal Statutory/Regulatory Incentives

Many health and environmental laws passed by Congress influence the environmental attributes of products (table 6-1).[1] Some, such as the Clean Air Act (CAA), Clean Water Act (CWA), and Resource Conservation and Recovery Act (RCRA) do so indirectly, by raising industry's costs of releasing wastes to the air, water, and land. Others, such as the Toxic Substances Control Act (TSCA) and the Federal Insecticide, Fungicide, and Rodenticide Act (FIFRA), control the use of hazardous chemicals and pesticides directly.[2]

Sometimes, design changes have resulted from "sunshine" laws that simply require the public disclosure of information about industry's use of toxic chemicals. For example, Title III of the Superfund Amendments and Reauthorization Act of 1986 requires manufacturers to report environ-

mental releases of 322 listed chemicals to a public database managed by the Environmental Protection Agency (EPA) called the Toxics Release Inventory (TRI). In several cases, the prospect of public disclosure of these releases stimulated companies to switch to more environmentally sound processes and product formulations.[3] Companies are also reformulating products to reduce potential liability for improper waste disposal under RCRA and the Comprehensive Environmental Response, Compensation, and Liability Act (CERCLA, or "Superfund").[4]

Of those laws listed in table 6-1, the Clean Air Act Amendments of 1990 may have the largest impact on product design, since they will result in restrictions on volatile organic compounds (Title I), hazardous air pollutants (Title III) and chlorofluorocarbons (CFCs) and other ozone-depleters (Title VI). These chemicals are used widely in manufacturing processes, as well as in paints, coatings, cleaners, pesticides, and household products.[5]

In listing the environmental laws in table 6-1, OTA does not intend to suggest that all environmental impacts of products are already regulated or that existing regulations provide adequate protection for the environment. Rather, the intent is to show the range of Federal laws that already affect product design.

Federal Disincentives?

Critics charge that some Federal regulations provide disincentives to green design. Examples often cited are government procurement policies (e.g., military specifications that require the use of virgin materials, CFC cleaners, and leaded paints where these materials are not necessary for product performance), RCRA regulations that make the

---

[1] For an overview of the influence of Federal laws on the formulation of various chemical products, see Kerr and Associates, Inc., "Effect of Environmental Statutory/Regulatory Requirements on Product Formulation/Process Design: Information on Solvents, Agricultural Chemicals, Products Containing Heavy Metals, and Related Household Cleaning Products," contractor report prepared for the Office of Technology Assessment, April 1992.

[2] Paul R. Portney (ed.), *Public Policies for Environmental Protection* (Washington, DC: Resources for the Future, 1990).

[3] For a discussion of how TRI reporting requirements changed the corporate culture at Monsanto and other companies, see Bruce Smart (ed.), *Beyond Compliance: A New Industry View of the Environment* (Washington, DC: World Resources Institute, 1992), p. 87.

[4] Kerr and Associates, op. cit., footnote 1.

[5] As one example, the total U.S. market for coatings in 1990 was $11.9 billion, but only about half of these coatings meet current environmental regulations for volatile organic compounds. Cited in promotional literature for a report by Business Communications Company, Inc. Norwalk, CT, "Environmentally Acceptable Coatings: The Industry," LC-136, May 1991.

Table 6-1—Federal Health and Environmental Laws Affecting Product Design

Statute	Impact on design	Agency
Clean Air Act of 1970 (and Amendments of 1977 and 1990)	Encourages reduction in the use of solvents, volatile organic compounds, and phases out chlorofluorocarbons.	EPA
Clean Water Act of 1977 (and Amendments of 1987)	Encourages reduction in the use of toxic chemicals that become water pollutants.	EPA
Resource Conservation and Recovery Act of 1976 (and Hazardous and Solid Waste Amendments of 1984)	Encourages redesign of products and processes to reduce generation of hazardous solvent, pesticide, and metal-bearing wastes, and to avoid liability for cleanup of wastes improperly disposed.	EPA
Comprehensive Environmental Response, Compensation, and Liability Act of 1980 (and Superfund Amendments and Reauthorization Act of 1986)	Encourages reduction in use of listed hazardous substances to avoid reporting requirements for releases of these substances, and liability for cleanup of Superfund sites.	EPA
Federal Insecticide, Fungicide, and Rodenticide Act of 1972 (and Amendments of 1988)	Encourages reformulation of pesticides to ensure safety and efficacy of active ingredients (and to avoid inert ingredients of toxicological concern), through a registration program.	EPA
Toxic Substances Control Act of 1976	Requires manufacturers to obtain approval from EPA (which may require submission of test data) before producing new chemicals that may pose an unreasonable risk to human health or the environment.	EPA
Federal Food, Drug, and Cosmetics Act	Regulates allowable pesticide residues in food, as well as the formulation of various solvent-containing cosmetic products.	FDA
Consumer Products Safety Act of 1978, Federal Hazardous Substances Act, Poison Prevention Packaging Act of 1970	Regulate the use of hazardous substances in consumer products.	CPSC
Occupational Safety and Health Act of 1970	Encourages manufacturers to avoid use of materials or processes that might expose workers to hazardous substances in the workplace.	OSHA

KEY: CPSC—Consumer Product Safety Commission; EPA—Environmental Protection Agency; FDA—Food and Drug Administration; OSHA—Occupational Safety and Health Administration.

SOURCE: Kerr and Associates, "Effect of Environmental Statutory/Regulatory Requirements on Product Formulation/Process Design: Information on Solvents, Agricultural Chemicals, Products Containing Heavy Metals, and Related Household Cleaning Products," a contractor report prepared for the Office of Technology Assessment, April 1992.

recycling of hazardous wastes more costly than disposing of them,[6] and the failure of RCRA regulations to distinguish between high-risk and low-risk chemicals and waste streams.[7] OTA did not attempt to evaluate these claims in this study, but Congress may wish to initiate further research in this area.

Several recent initiatives could help to remove some of the barriers to green design that exist in current Federal rules and regulations. In October 1991, President Bush signed Executive Order 12780,

the Federal Recycling and Procurement Policy, which requires Federal agencies to increase recycling and waste reduction efforts and to encourage markets for recovered materials by favoring the purchase of products with recycled content.[8] The order creates a Federal recycling coordinator and a Council on Federal Recycling and Procurement. It also requires each agency to designate its own recycling coordinator. Recently, the Department of Defense issued directives emphasizing waste prevention through the acquisition process and through

---

[6] See testimony of Herschel Cutler, Institute of Scrap Recycling Industries, before the Subcommittee on Environmental Protection of the Senate Committee on Environment and Public Works, June 5, 1991.

[7] James Bovard, "RCRA: Origin of an Environmental Debacle," *Journal of Regulation and Social Costs*, vol. 1, No. 2, January 1991, p. 37.

[8] *Environmental Quality 1991, 22nd Annual Report of the Council on Environmental Quality* (Washington, DC: U.S. Government Printing Office, March 1992), p. 113.

military specifications and standards. Some 40,000 military specifications requiring the use of hazardous materials are currently under review.[9] [10] These initiatives could help to stimulate market demand for green products.

State and Local Laws and Regulations

Many State and local governments are also enacting policies aimed at reducing the environmental impacts of products (table 6-2). These measures include mandatory industry plans to reduce the use of toxic chemicals, requirements for industry disclosure of hazardous chemicals in products, and creation of standard definitions for advertisers' use of environmental terms such as ''recycled.'' States have also enacted some targeted product control measures such as recycled content requirements for newspaper, bans and taxes on specific packages, mandated manufacturer takeback of batteries, and tax incentives for recycling. In some cases, these laws regulate products and processes more strictly than do Federal laws. Notable examples are California's regulations on auto emissions, permissible volatile organic compound content of products, and labeling requirements for products containing carcinogens and reproductive toxics.[11]

The lack of uniform Federal environmental standards for products is alarming to industry, which fears having to satisfy different regulations in each State.[12] [13] This prospect is especially of concern for products that are distributed through national networks. Companies are faced with the choice of redesigning products to meet the most stringent State requirement, or changing their distribution systems. OTA did not evaluate these concerns in this study, but Congress may wish to investigate further the extent to which the diversity of State regulations may impose unnecessary additional costs on indus-

try, and where Federal intervention may be appropriate to establish national guidelines for environmental product policy (see section on coordination and harmonization below).

Market Incentives

Manufacturers already have a number of economic incentives to move toward green design. By reducing the quantity of materials used in products, they can reduce their manufacturing costs. This incentive partially accounts for the trend toward increasing efficiency of materials use described in chapter 2. Manufacturer's waste disposal costs are also increasing as permitted landfill capacity continues to shrink and waste is shipped greater distances for disposal. This provides an incentive for waste prevention and in-process recycling of scrap.

There are also marketing opportunities to gain the loyalty of environmentally conscious consumers. Surveys indicate that consumer interest in the environmental attributes of products is on the rise, and that a substantial segment is willing to pay a premium for environmentally sound products.[14] Environmental regulations are also creating new market opportunities for small firms with innovative environmental technologies.[15]

Corporate Responses

Manufacturers are responding to these incentives in many ways. For instance, less toxic substitutes for heavy metals have been adopted in such products as inks, paints, plastics, and batteries; the electronics industry has redesigned its manufacturing processes to drastically reduce the use of CFCs; and several companies are redesigning products and packaging to be lighter, more compostable, or to use recovered materials.[16] Environmental advertising is now being used to sell a broad range of products, from gasoline to fabric softener.[17] A growing number of companies

[9] Ibid., p. 157.

[10] See testimony of David J. Berteau, Principal Deputy Assistant Secretary of Defense (Production and Logistics) before the Subcommittee on Oversight of Government Management of the Senate Committee on Governmental Affairs, Nov. 8, 1991.

[11] See Kerr and Associates, op. cit., footnote 1, and Paul R. Portney, op. cit., footnote 2, p. 282.

[12] John Holusha, ''States Lead on Environment and Industries Complain,'' *The New York Times,* Apr. 1, 1991, p. D1.

[13] Gary D. Sesser, ''Just Who's in Charge Here?'' *Across the Board,* July/August 1991, p. 11.

[14] See, e.g., The Roper Organization, Inc., ''The Environment: Public Attitudes and Individual Behavior,'' a study conducted for S.C. Johnson and Son, Inc., July 1990.

[15] Mark Fischetti, ''Green Entrepreneurs,'' *Technology Review,* April 1992, p. 39.

[16] See, e.g., Bruce Smart (ed.), op. cit., footnote 3.

[17] See, e.g., ''Selling Green,'' *Consumer Reports,* October 1991, p. 687.

Table 6-2—Examples of State or Local Laws Affecting Product Design

Provision	State	Comments
Packaging		
Ban on multilayered aseptic beverage containers.	Maine	No other States have followed Maine's example.
Ban on polystyrene-foam food packaging.	Minneapolis/St. Paul, MN and Portland, OR	These local bans, which have not been enforced, are giving way to recycling mandates.
Ban on the use of toxic heavy metals in packaging.	10 States	These laws are based on model legislation developed by the Coalition of Northeastern Governors.
Volatile organic compounds		
Mandatory reductions in VOC content in consumer products.	California	Reductions may require reformulation costing $100,000 to $2 million per product.
Environmental labeling		
Regulations on the use of environmental terms, such as "recyclable" or "recycled."	13 States	States vary in the requirements that a product must meet to qualify for use of environmental terms and symbols.
Labeling requirements for products that contain chemicals listed as carcinogenic or causing birth defects.	California	The list of chemicals differs substantially from Federal lists. Products must be labeled even if listed chemicals are present in trace amounts.
Newsprint		
Recycled content requirements for newspapers.	10 States	These requirements have driven substantial industry investment in newsprint recycling equipment.
Batteries		
Limits on mercury in household batteries.	4 States	Mercury has largely been removed from household batteries in recent years.
Requirements for manufacturers to take back and recycle rechargeable batteries.	4 States	Rechargeable batteries currently contain the toxic heavy metals nickel and cadmium.
Requirements for all batteries to be "easily removable" from products.	Connecticut and New York	May require significant design changes.
Toxic use reduction		
Requirements for companies to submit plans for reducing their use of listed toxic chemicals.	5 States	Involves "voluntary" industry goal-setting with public disclosure of progress toward the goals.

SOURCE: Office of Technology Assessment, 1992.

are participating in waste exchanges, where one company's waste becomes another's raw material.[18]

Recognizing that customers and government regulators will be paying greater attention to the environmental attributes of products in the future, numerous industry trade associations, professional engineering and design societies, and consortia are addressing these issues (table 6-3). Activities of these organizations include programs to promote "product stewardship" (manufacturer responsibility beyond the factory gate), standards for labeling of recyclable materials, design concepts for product disassembly, etc.

The existence of these private organizations does not necessarily mean that the participating companies have all taken the environmentalists' agenda to heart. On the contrary, some are participating for defensive reasons, to promote the environmental

benefits of current materials and products, or to lobby against new environmental regulations. Nevertheless, the existence of these programs is evidence that the companies believe that increased environmental scrutiny of products and processes is inevitable, and that they are better off taking the initiative rather than merely reacting. In the future, these industry organizations could provide useful forums for information exchange on green design.

Response of Educational Institutions

Although the concept of green design was articulated more than 20 years ago, it has not been integrated into the education and training of designers, engineers, and business managers. A recent survey by the EPA's National Advisory Council on Environmental Policy and Technology found that only 10 to 15 of the nearly 400 engineering schools

---

[18] Rodney Ho, "Waste Exchanges Help More Companies Bag a Treasure From Another's Trash," *Wall Street Journal*, Aug. 2, 1991, p. A5B.

Table 6-3—Industry and Professional Organizations Concerned With Green Design

Organization	Activities	Comments
Industry Trade Associations and Coalitions[a]		
American Electronics Association/Task Force on Design for the Environment	Holds regular meetings to share information on activities in the member companies, and to develop strategies for green design.	Membership includes the major electronics and computer companies, as well as representatives from aerospace and automotive industries.
Chemical Manufacturers Association (CMA)	Initiated the Responsible Care Program, a code of management practices developing the idea of "product stewardship," which extends company responsibility for a product beyond the factory gate.	Membership includes major manufacturers of chemical products.
Chemical Specialties Manufacturers Association (CSMA)	Promotes waste prevention activities such as product reformulation or process modification. Provides information through brochures and conferences to educate membership and consumers on proper use, storage, recycling, and disposal of products.	Represents companies engaged in the formulation, manufacture, packaging, marketing, and distribution of products to households, institutions, and industries. Membership includes 80 percent of the domestic aerosol industry production capacity.
Council on Plastics and Packaging in the Environment (COPPE)	Promotes waste prevention and recycling of plastic packaging. Sponsors meetings and provides information on plastic packaging and solid waste issues.	Coalition of plastic resin producers, packaging manufacturers and users, and trade associations.
Global Environmental Management Initiative (GEMI)	Promotes a worldwide environmental ethic in business management. Sponsors conferences examining the connections between product design, total quality management, and environmental excellence.	Coalition of 22 leading companies including chemical and consumer product manufacturers.
Institute of Scrap Recycling Industries (ISRI)	Promotes design for recyclability and removal of hazardous materials from products	Represents 1,800 firms involved in all major recycled commodities.
National Paint and Coatings Association (NPCA)	Paint Pollution Prevention Program aims to reduce environmental impacts of paints through efficient material utilization, toxic use reduction, and product stewardship.	National umbrella group for regional paint and coatings associations.
Society of the Plastics Industry (SPI)/Partnership for Plastics Progress (PPP)	Promotes plastic recycling programs. Has developed a labeling system to identify plastics by resin type to facilitate separation for recycling. Is developing design strategies for better management of plastics used in durable goods.	PPP is a task force representing major companies in the plastics industry.
Vehicle Recycling Partnership (VRP)	Sponsors meetings and funds research on methods to enhance auto recycling and better management of materials through better design.	Consists of Ford, Chrysler, and General Motors, as well as materials suppliers, dismantlers, and recyclers.

(continued on next page)

Table 6-3—Industry and Professional Organizations Concerned With Green Design—Continued

Organization	Activities	Comments
Professional Societies		
American Institute of Architects (AIA)	Developing environmentally sound approaches to the design of buildings. With funding from EPA, is developing an Environmental Resource Guide, containing environmental information on building materials and case studies in green design.	Professional society representing architects.
American Institute of Chemical Engineers (AIChE)	Established the Center for Waste Reduction Technologies, whose goal is to integrate the design of production facilities with waste management requirements. The Center is an umbrella organization to conduct research and education with funding from government, universities, and industry.	Professional society of chemical engineers. Industry sponsors of the Center include chemical companies, manufacturers, and engineering services firms.
Industrial Designers Society of America (IDSA)	Sponsors conferences on product design and the environment, and has devoted several issues of its journal *Innovation* to discussions of green design.	Professional society representing industrial designers.
Institute of Packaging Professionals (IoPP)	Issued "Packaging Reduction, Recycling, and Disposal Guidelines."	Organization for packaging professionals for consumer, industrial, and military products.
Society of Environmental Toxicology and Chemistry (SETAC)	Sponsored meetings to develop a technical framework for life-cycle assessments.	Professional society of 2,000 members that provides a forum on resource use issues for environmental scientists and engineers from academia, industry, government, and public interest groups.
Society for the Advancement of Material and Process Engineering (SAMPE); Society of Plastics Engineers (SPE); American Society of Mechanical Engineers (ASME); Society of Automotive Engineers (SAE)	All have sponsored conferences on issues related to green design.	Professional societies representing various engineering specialties.

a In addition to these purely industry organizations, there are also several hybrid organizations in which industry groups work together with government or environmental groups to promote environmentally sound business practices. Examples include the Coalition of Northeastern Governors (CONEG) Source Reduction Council, which has developed "preferred packaging guidelines," and the Conservation Foundation/World Wildlife Fund, which has developed environmental curricula for use in business schools.

SOURCE: Office of Technology Assessment, 1992.

in the United States offer significant coursework in waste prevention.[19] The Management Institute of Environment and Business has estimated that only about 25 of the 700 schools of management and business have a course on business and the environment, and none requires the course for graduation.[20]

This is beginning to change, though: OTA found that interest in this topic among design, architecture, engineering, and business schools is high, and several schools have begun to integrate environmental issues into their curricula (see table 6-4).

This may be an opportune time to inject an environmental dimension into the educational experience of designers and engineers. Many schools are reevaluating their curricula in light of growing criticism that students are not being prepared in the "best practice" design techniques used by the most competitive companies.[21] [22] This reevaluation process could provide a window of opportunity to add environmental courses or projects.

As part of this study, the American Society for Engineering Education's Engineering Deans Council conducted an informal survey of Engineering School Deans on behalf of OTA (see box 6-A). Fifteen of the twenty respondents indicated that their institution already offers some sort of environmental program for students. In most cases, however, these programs take the form of optional classes on pollution control or "environmental engineering." Only five respondents reported that they are actively incorporating environmental concerns into their standard engineering courses. Lack of funding and

lack of faculty training were cited as significant barriers to further progress.

These needs are beginning to be addressed by both Federal and private programs. EPA has funded the National Pollution Prevention Center at the University of Michigan, which is developing waste prevention curriculum materials for colleges and universities, including modules for industrial design and engineering design courses.[23] The center also plans to provide information and education to university faculty through interdepartmental seminars. In another example, the National Wildlife Federation's Corporate Conservation Council has sponsored a pilot program to introduce environmental issues into business school education.[24] [25]

Information Resources Available

A variety of information resources relevant to green design are becoming available. These include books that offer general guidelines for green design,[26] as well as information more appropriate for specific industries and manufacturing processes.[27] Experimental computer programs are being developed to assist designers in evaluating their choices according to life-cycle criteria.[28] Information is also available on a variety of related topics, such as how to evaluate design decisions by total cost assessment[29] and how to conduct materials balance assessments and waste stream audits.[30] [31]

Electronic Networks

Electronic networks can provide useful forums for information exchange among those interested in

---

[19] Anthony D. Cortese, "Education for an Environmentally Sustainable Future," *Environmental Science and Technology*, vol. 26, No. 6, 1992, p. 1108.

[20] Ibid.

[21] John R. Dixon, "New Goals for Engineering Education," *Mechanical Engineering*, March 1991, p. 56.

[22] National Research Council, *Improving Engineering Design: Designing for Competitive Advantage* (Washington, DC: National Academy Press, 1991), p. 35.

[23] Environmental Protection Agency, "Pollution Prevention Resources and Training Opportunities in 1992," EPA/560/8-92-002, January 1992, p. 94.

[24] James E. Post, "The Greening of Management," *Issues in Science and Technology*, summer 1990, p. 68.

[25] At this writing, the Education and Training Committee of EPA's National Advisory Council for Environmental Policy and Technology was completing work on a national strategy to encourage waste prevention education and training.

[26] See, e.g., Dorothy Mackenzie, *Design for the Environment* (New York, NY: Rizzoli, 1991), and references cited therein.

[27] Environmental Protection Agency, "Pollution Prevention Resources and Training Opportunities in 1992," op. cit., footnote 23, p. 13.

[28] The software, called Simapro, is available from PRé Consultants, Amersfoort, The Netherlands.

[29] Environmental Protection Agency, "Total Cost Assessment: Accelerating Industrial Pollution Prevention Through Innovative Project Financial Analysis," a report prepared by Tellus Institute, Boston, MA, May 1992.

[30] Environmental Protection Agency, "Facility Pollution Prevention Guide," EPA/600/R-92/088, May 1992.

[31] Lauren Kenworthy and Eric Schaeffer, "A Citizen's Guide to Promoting Toxic Waste Reduction," published by Inform, Inc., 1990.

Table 6-4—Environmental Education Programs in Design, Engineering, and Business Schools

Institution	Activity
Boston University	Offers graduate business courses on managing environmental issues.
Carnegie Mellon University	A variety of courses and seminars are being developed around the idea of design for the environment, involving the Center for Solid Waste Management Research, the Environmental Institute, and the Engineering Design Research Center.
Grand Valley State University	The Waste Reduction and Management Program is developing engineering curricular materials on green design and provides seminars for engineers and faculty on "cutting edge" design approaches.
Loyola University	Offers graduate business courses on managing environmental issues.
Massachusetts Institute of Technology	The Technology, Business, and the Environment Group offers workshops and seminars for engineers and managers, and works to integrate waste prevention concepts into undergraduate and graduate courses.
Rhode Island School of Design	Incorporates environmental concerns into the industrial design curriculum through course material and projects.
Tufts University	The Tufts Environmental Literacy Institute seeks to incorporate environmental concerns throughout the curriculum; the Center for Environmental Management provides education and training programs for engineering students.
University of the Arts (Philadelphia)	Incorporates environmental concerns into industrial design curriculum through course material and class projects.
University of California at Los Angeles	Integrates environmental concerns throughout engineering disciplines through problem sets and projects.
University of Michigan	The Pollution Prevention Center for Curriculum Development and Dissemination is developing curriculum modules for undergraduate and graduate courses in engineering, business, and science. Summer workshops and seminars are also offered.
University of Minnesota	Offers graduate business courses on managing environmental issues.
University of Rhode Island	Students in the Chemical Engineering Department evaluate waste prevention opportunities for Rhode Island firms.
University of Wisconsin	The Engineering Professional Development Program offers short courses to engineering students on waste prevention and green design.

SOURCE: Office of Technology Assessment, 1992.

green product design. One existing government-funded network is EPA's Pollution Prevention Information Exchange (PIES), which is part of the Pollution Prevention Information Clearinghouse.[32] PIES contains bibliographic materials, industry case studies, announcements, and an electronic bulletin board that allows users to send messages to one another. Although PIES was not organized specifically with the needs of designers in mind, it contains a considerable amount of relevant information, and in the future it could be expanded to include comparative environmental information on alternative materials, substitutes for toxic chemicals, etc.

Consumer Information

One of the most powerful determinants of product design is consumer preference. Yet unless consumers are able to recognize green products in the store, this potentially powerful incentive for green design is neutralized. To fill this need, several "green consumer guides" have become available in recent years,[33] though some of the recommendations in these guides have been controversial.[34]

It is especially important to reach the next generation of consumers early. As authorized by the National Environmental Education Act of 1990,

[32] Environmental Protection Agency, "Pollution Prevention Resources and Training Opportunities in 1992," op. cit., footnote 23, p. 100.

[33] See, e.g., Joel Makower, John Elkington, and Julia Hailes, *The Green Consumer*, Penguin Books, New York, 1990; The Earthworks Group, *50 Simple Things You Can Do To Save the Earth* (Berkeley, CA: Earthworks Press, 1989); Debra Lynn Dadd, "Nontoxic and Natural," Jeremy P. Tarcher, Los Angeles, 1984.

[34] EPA withdrew copies of its publication "The Environmental Consumer's Handbook" (EPA/530-SW-90-034B, October, 1990) after industry protested the implication in the report that disposable or multi-material products are environmentally less desirable.

Box 6-A—Survey of Engineering School Deans

On behalf of OTA, the American Society for Engineering Education's Engineering Deans Council undertook a survey of the views of Engineering Deans regarding the need to integrate environmental concerns into engineering school curricula. Of the 20 respondents, 17 considered such integration to be very important, and 16 believed that environmental courses or programs would help attract new engineering students.

When the Deans were asked to comment on how their institutions were addressing this issue, 15 cited some form of ongoing environmental program, and several more cited programs being planned. The most common approach was to offer optional courses on environmental topics within chemical, mechanical, or civil engineering programs (11 schools). Seven schools reported that new majors or degree programs were being developed (typically in "Environmental Engineering"). Only 5 schools indicated that they are integrating environmental concerns into their standard engineering courses through modules, projects, or problem sets.

The Deans were also asked to comment on what barriers exist to incorporating a stronger environmental perspective into engineering programs. The two most frequently cited responses were the lack of money and the availability of appropriately trained faculty (each cited by five respondents). Other answers included a lack of course materials, and a curriculum already crammed with other topics.

When asked what the Federal Government could do to help engineering schools incorporate environmental concerns into engineering education and research, 10 of the respondents indicated that more Federal funding was necessary (research funds, scholarships, training). Seven indicated that Federal assistance with curriculum development, course development, or new programs would be beneficial. Other suggestions included having the Federal Government increase national awareness and concern regarding green design, to establish new research centers through the National Science Foundation, to create a competitive award to highlight work in this area, and to identify more clearly the Nation's most pressing environmental problems.

SOURCE: ASEE Engineering Deans Council Survey for OTA.

EPA has established an Office of Environmental Education to

> . . . foster an enhanced environmental ethic in society by improving the environmental literacy of our youth and increasing the public's awareness of environmental problems.

The primary focus will be on grade levels K-12. EPA has also established an agencywide National Pollution Prevention Environmental Education Task Force to develop educational materials for students and teachers in grades K-12.[35]

Ongoing Activities

Finally, OTA identified a number of ongoing Federal activities that will provide additional information for designers in the near future (table 6-5). EPA is the agency most directly involved; for example, EPA's Office of Research and Development is supporting the development of a Life Cycle Design Guidance Manual, which is intended to explore how designers can incorporate life cycle assessment into their designs. EPA also has a project underway with the American Institute of Architects to develop an Environmental Resource Guide, to assist architects in making environmentally sound choices of construction materials. Table 6-5 also identifies several relevant projects in other agencies, including the Department of Energy and the National Science Foundation.

In the Pollution Prevention Act of 1990, Congress required that manufacturers who report their releases of toxic chemicals for the TRI must also report how these releases were affected by waste prevention activities, including product and process redesign. When these data become available (probably some time in 1993), they could provide valuable insight into an area where little information currently exists: how product design choices affect industrial waste streams.

CONGRESSIONAL ROLE

These ongoing activities suggest that green design is a concept that is gathering momentum. Even if Congress takes no further action, the incentives

---

[35] Environmental Protection Agency, Pollution Prevention Resources and Training Opportunities in 1992, op. cit., footnote 23, p. 95. See also Environmental Protection Agency, "Environmental Education Materials for Teachers and Young People (Grades K-12)," 21K-1009, July 1991.

Table 6-5—Federally Funded Programs Related to Green Design

Agency/Office	Program/activity	Comments
Department of Energy		
Office of Industrial Technologies	Industrial Waste Reduction Program	A research and development program to identify priority industrial waste streams, assess opportunities for addressing these waste streams through redesigning products and production processes, and assess technology transfer from national laboratories.
Environmental Protection Agency		
Office of Research and Development	Environmental Resource Guide	Contracted to the American Institute of Architects, this project will provide information to architects on the life cycle environmental impacts of construction materials.
	Life Cycle Assessment Methodology	Contracted to Battelle and Franklin Associates, Ltd., this project will develop standard methodologies for conducting product life-cycle assessments.
	Clean Products Case Studies	Contracted to INFORM Inc., this project will provide case studies of green design, especially the reduced use of toxic substances in products.
	Safe Substitutes	Contracted to the University of Tennessee, this project will identify priority toxic chemicals and evaluate possible substitutes.
	Life Cycle Design Guidance Manual: Environmental Requirements and the Product System	Contracted to the University of Michigan, this manual will explore how designers can incorporate life-cycle information in their designs.
	National Pollution Prevention Center	Located at the University of Michigan, this center is developing waste prevention information modules for industrial and engineering design courses.
	American Institute for Pollution Prevention	In association with the University of Cincinnati, the institute serves as a liaison to a broad cross-section of industry, with projects involving four aspects of waste prevention: education, economics, implementation, and technology.
Office of Pollution Prevention and Toxics	Design for the Environment	Scheduled to be launched in September 1992, this program will gather, coordinate, and disseminate information on green design.
National Science Foundation	Engineering Design Research Center	Located at Carnegie Mellon University, the center is organizing a program to explore methods for green design.

SOURCE: Office of Technology Assessment, 1992.

discussed above can be expected to continue in the future. Implementation of tougher emissions standards under the Clean Air Act Amendments of 1990 will increase pressures on companies to reduce their use of hazardous solvents and other volatile organic compounds. New regulations requiring liners and leachate collection systems in landfill construction will increase the costs of solid waste disposal and provide increased incentives for waste prevention.[36] Various States will no doubt continue to pass legislation to regulate the environmental attributes of products and waste streams. And as consumers become more attuned to environmental concerns, they will increasingly demand that manufacturers take more responsibility for the environmental impacts of their products.

Despite these incentives, though, OTA finds there are four areas where congressional action is needed to maintain existing momentum and foster further progress:

- *Research.* At present, designers and policymakers don't know what materials or waste streams are of greatest concern, or how product designs might be changed to address them most effectively. Private companies have little incentive to conduct this research.
- *Credible information for consumers.* Surveys show that consumers are interested in green products, but most don't know what is "green." As discussed in chapters 3 and 4, defining what's green is a multidimensional problem. In the absence of Federal action to establish consistent ground rules defining terms and measurement methods, the growing interest of consumers could become dissipated in confusion and skepticism.
- *Market distortions and environmental externalities.* Most observers agree that the prices of materials and energy do not reflect their true environmental costs. Failure to internalize these environmental costs into design and production decisions can make environmentally sound choices seem economically unattractive. Further, they argue that some government policies, such as subsidies for the extraction of virgin materials, also distort prices.
- *Coordination and harmonization.* OTA found that several research projects related to green

design are being sponsored by various Federal agencies and offices (table 6-5), but that there is little or no coordination among them. And unlike its major competitors, the United States has no institutional focus at the Federal level for addressing environmental product policy (see chapter 5).

These issues are discussed in greater detail below, and options for addressing them are presented. The chapter concludes with a short list of relatively quick and inexpensive options Congress could choose to encourage green design.

RESEARCH NEEDS

To take full advantage of the potential of green design, both designers and policymakers need more information about where the major opportunities lie and the most cost-effective ways of addressing them. This information must be developed through research. Below, OTA discusses research needs in two categories: technical research and applied social science research. In this study, OTA has made no attempt to evaluate how much additional funding may be necessary to fully address these needs.

Technical Research

Setting Priorities Based on Risk

Beyond certain obvious imperatives such as avoiding the use of CFCs, designers and policymakers have little information as to what materials and waste streams pose the greatest health and environmental risks. Current lists of hazardous substances subject to various State and Federal regulations contain hundreds of chemicals, each having different uses and posing different risks to health and the environment.

Several research efforts funded by the Department of Energy (DOE) and EPA are attempting to identify priority products and waste streams (see table 6-5). **Congress could require EPA and DOE to jointly identify a short list of products and production processes that appear to pose the greatest health and environmental risks.** This would be consistent with EPA's stated goal of reevaluating its priorities

---

[36] EPA's final rule specifying minimum Federal criteria for municipal solid waste landfill design was published in the *Federal Register*, vol. 56, No. 196, Oct. 9, 1991, and becomes effective on October 9, 1993.

based on risk assessment.[37][38] In the case of chemicals, a possible starting point might be the list of 17 categories of chemicals that EPA has targeted in its 33/50 Program.[39] Armed with this information, designers can find appropriate substitutes and avoid dissipative uses of these materials.

Safe Substitutes

To reduce overall environmental risks, the risk tradeoffs of switching from one chemical to another must be understood. Currently, designers may be substituting regulated chemicals of relatively known risk with unregulated chemicals of unknown risk. The assumption is often made that the new chemicals are safer, but this may not be the case. Although the Toxic Substances Control Act of 1976 requires EPA to consider the net risk of chemical substitution in regulating chemicals, most chemicals have been tested on a one-at-a-time basis.[40] **Congress could direct EPA to evaluate the risks of the priority chemicals in "use clusters"—groups of chemicals that can substitute for one another (e.g., solvents or coolants).[41]** This comparative information could then be made available to designers through such mechanisms as EPA's Pollution Prevention Information Exchange.

Understanding Materials Flows

Policymakers need better models of how various materials and wastes of concern flow through the economy and into the environment. These models can help identify the major sources of environmental pollutants and the most cost-effective ways of reducing them. Without this information, resources may be diverted to address the most visible problems, rather than the most serious ones. For instance, 10 States have banned the use of toxic heavy metals in packaging, yet this source contributes only a few percent of heavy metals in landfills and incinerators.[42][43] As another example, there has been considerable concern expressed about the release of mercury from the incineration of municipal solid waste, yet these releases may be small compared with mercury releases from coal combustion in power plants.[44]

These examples underline the need for detailed "materials balance" analyses that quantitatively track materials of special concern through initial production, use in industrial processes and products, and disposal.[45] Preliminary materials balance studies have been carried out for several hazardous substances,[46][47] but a more systematic approach is needed. **For the short list of high-risk materials identified above, Congress could direct EPA and**

---

[37] William Reilly, "Taking Aim Toward 2000: Rethinking the Nation's Environmental Agenda," *Environmental Law*, vol. 21, No. 4, 1991, p. 1359. For a discussion of the limitations of risk reduction as a strategy for the future, see also John Atcheson, "The Department of Risk Reduction or Risky Business," ibid., p. 1375.

[38] Techniques for evaluating environmental risks are still evolving. EPA's Risk Assessment Forum is sponsoring several case studies that could help to establish a framework for ecological risk assessment. Preliminary results are expected in 1994. See "Environmental Agency Launches a Study in Ecological Risk Assessment," *Science*, Mar. 20, 1992, p. 1499.

[39] The goals of the 33/50 Program are to reduce industry releases of the 17 target chemicals 33 percent by 1993 and 50 percent by 1995, based on 1988 levels. In selecting these chemicals, EPA started with the list of 322 TRI chemicals and employed a screening process based on volume of production, volume of releases, and hazardous properties. See Environmental Protection Agency, "Pollution Prevention Resources and Training Opportunities in 1992," op. cit., footnote 23, p. 84.

[40] Michael Shapiro, "Toxic Substances Policy," in Paul R. Portney, op. cit., footnote 2, p. 224.

[41] EPA's Office of Pollution Prevention and Toxics has announced plans to employ the "use cluster" concept for evaluating substitutes for hazardous chemicals in the future. Jean E. Parker, Office of Pollution Prevention and Toxics, personal communication, August 1992.

[42] These laws are based on model legislation developed by the Source Reduction Council of the Coalition of Northeastern Governors (CONEG).

[43] The dominant source of lead and cadmium in municipal solid waste is in batteries, especially lead-acid automobile batteries. Some 38 States now have laws regulating the disposal of batteries. Another major source is consumer electronics, whose disposal is generally not regulated. See Franklin Associates, Ltd., "Characterization of Products Containing Lead and Cadmium in Municipal Solid Waste in the United States, 1970-2000," a report prepared for the Environmental Protection Agency, January 1989.

[44] According to one estimate, about 65 percent of anthropogenic mercury emissions to the atmosphere is due to coal burning, and another 25 percent is due to waste incineration. F. Slemr and E. Langer, "Increase in Global Atmospheric Concentrations of Mercury Inferred From Measurements Over the Atlantic Ocean," *Nature*, vol. 355, Jan. 30, 1992, p. 436.

[45] This is not a new idea, but it has never been pursued systematically. See Allen V. Kneese, Robert U. Ayres, and Ralph C. d'Arge, "Economics and the Environment: A Materials Balance Approach," a monograph published by Resources for the Future, Washington, DC, 1970.

[46] R.U. Ayres et al., "Industrial Metabolism, the Environment, and Application of Materials-Balance Principles for Selected Chemicals," International Institute for Applied Systems Analysis, RR-89-11, Laxenburg, Austria, 1989.

[47] David T. Allen, "Wastes as Raw Materials," presented at the National Academy of Sciences Workshop on Industrial Ecology/Design for the Environment, Woods Hole, MA, July 16, 1992.

DOE to conduct detailed materials balance studies showing how these materials flow through the economy and into the environment.

A significant barrier to better modeling of materials flows is the quality of data on industrial waste streams. More than 20 national sources of data are available (e.g., the TRI, or the biennial survey of hazardous waste generators required by RCRA).[48] These databases, which were established for different purposes, cover different waste generators, waste types, and time periods. This makes it difficult to get a coherent picture of materials flows, whether one is interested in tracking specific materials or the performance of specific industrial sectors. **Congress could direct EPA, DOE, and the Department of Commerce (DOC) to jointly explore how existing waste stream reporting requirements might be harmonized to provide a more coherent picture of waste flows.[49]**

A Scientific Basis for Better Materials Management

OTA found few examples of research relating to better materials management at the Federal level. Instead, most Federal research projects relating to green design appear to be oriented toward preventing the release of toxic or hazardous materials (see, for instance, table 6-5). This is also reflected in the solid waste dichotomy defined by RCRA, in which "hazardous" solid waste is regulated by the Federal Government, while responsibility for "nonhazardous" solid waste management is delegated to the States.

To improve the connectivity between product design and waste management, Congress could establish a grant program for joint research and demonstration projects having both a design component (e.g., to develop principles of design for remanufacturing, disassembly, compostability, etc.) and a waste management component (e.g., to develop improved recycling, composting, and incineration technologies). Examples of the fruits of this research might be adhesives, paints, or coatings that do not inhibit recycling processes; mixed materials that can be co-recycled without sacrificing the properties of the finished product; or materials that generate fewer toxic residues when incinerated. Materials derived from biological sources are another important category; this could lead to a class of renewable materials that might be extremely durable or fully biodegradable.

One mechanism for funding such joint projects may be the National Science Foundation's Engineering Research Centers.[50] Another avenue may be the Advanced Materials and Processing Program, an interagency materials research initiative for materials science and engineering announced by President Bush in 1992.[51]

Applied Social Science Research

A 1990 National Research Council workshop concluded that EPA's waste reduction research program emphasizes technical issues to the exclusion of applied social science research.[52] The workshop participants singled out three categories of special need: measurement techniques for evaluating progress, institutional and behavioral barriers, and the need for more analysis of policy incentives. OTA finds that these same research needs apply to green design, as discussed below.

Measurement Techniques

Measuring what is "green" is one of the most difficult challenges facing designers and policymakers. Designers must have targets for weight reduction, and substitution for toxic chemicals. Public interest groups need criteria against which to evaluate industry progress, and companies need criteria to be able to claim credit for legitimate environmental improvements.

---

[48] Jack Eisenhauer and Richard Cordes, "Industrial Waste Databases: A Simple Roadmap," *Hazardous Waste and Hazardous Materials*, vol. 9, No. 1, 1992, p. 1.

[49] Preliminary work along these lines is currently being funded by the Department of Energy's Office of Industrial Technologies. See Alan Schroeder, "Industrial Waste Sources in the U.S.A.," in the *Proceedings of Global Pollution Prevention—'91*, Washington, DC, Apr. 3-5, 1991, p. 229.

[50] The Engineering Design Research Center at Carnegie Mellon University is developing an industry consortium interested in exploring principles of design for disassembly and recycling.

[51] EPA has proposed a project on Environmentally Benign Materials and Processes as part of the fiscal year 1993 enhancements to the Advanced Materials Processing Program. See "Advanced Materials and Processing: the Fiscal Year 1993 Program," a report by the Federal Coordinating Council for Science, Engineering, and Technology's Committee on Industry and Technology, Office of Science and Technology Policy, April 1992.

[52] National Research Council, Committee on Opportunities in Applied Environmental Research and Development, *Waste Reduction: Research Needs in Applied Social Sciences, a Workshop Report* (Washington, DC: National Academy Press, 1990).

As discussed in chapter 4, green design always involves making tradeoffs. In principle, a comprehensive life-cycle analysis (LCA) of a product or process can indicate how to make these tradeoffs.[53] A challenge for the future is to develop streamlined LCA methods that focus on a few critical parameters. It may also be possible to develop narrower design rules of thumb or "green indicators" for use by designers for specific products or facilities (e.g., a tire design might be evaluated based on its expected service life divided by its weight).[54] These green indicators would be expected to vary for different products.

A central measurement issue for policymakers is how to measure the waste prevention attributes of a product. Whereas recycling rates or recycled content are *relatively* easy to measure, quantifying waste prevention is notoriously difficult.[55] [56] **Yet if waste prevention is indeed preferred to recycling in the solid waste management "hierarchy," as stated both by EPA and by Congress in the Pollution Prevention Act of 1990, it is important that mechanisms be found to credit waste prevention in government procurement programs and legislation that aim to increase recycling rates.**

Finally, to support system-oriented product design (see chapter 4), new macro-level metrics will be required that can characterize the environmental performance of alternative production and consumption systems (e.g., alternative ways of providing the same service), rather than just alternative *products*. Suggestions for such metrics include dematerialization (reductions in the weight of materials used to provide a given level of goods and services), decarbonization (reduction in the quantity of fossil fuels consumed to provide a given level of goods and services), and input-output analysis of production systems, accounting for both products and waste streams.[57]

Many of these measurement issues are controversial, and are probably best addressed through a consensus-building process involving government, industry, universities, and public interest groups. **Congress could provide funding to EPA to convene a series of consensus-building workshops involving all interested stakeholders to resolve these measurement issues.[58]**

Institutional and Behavioral Research

To explore the full potential of green design, a better understanding is needed of how companies manage the design function and how design decisions vis-a-vis the environment are affected by such factors as type of product, company size, and corporate culture. For example, barriers to green design arise from cost accounting procedures[59] and other institutional or behavioral factors. Research is also needed to understand how companies shape and are shaped by customer needs. In particular, this research could include how individual consumers and large-volume commercial buyers view environmental risks and make decisions to purchase environmentally preferred products (see below).

Congress could provide funding through EPA, the National Science Foundation (NSF), or the National Institute for Standards and Technology (NIST) for a series of industry case studies to analyze how institutional and behavioral factors influence design decisions vis-a-vis the environment in a variety of industry settings (including

---

[53] See the upcoming report, "Product Life-Cycle Assessment: Inventory Guidelines and Principles," a study prepared for EPA's Office of Research and Development by Battelle and Franklin Associates, Ltd.

[54] D. Navinchandra, "Design for Environmentability," in *Proceedings of the ASME Design Theory and Methodology Conference*, American Society of Mechanical Engineers, Miami, FL, 1991.

[55] U.S. Congress, Office of Technology Assessment, *Serious Reduction of Hazardous Waste: For Pollution Prevention and Industrial Efficiency*, OTA-ITE-317 (Washington, DC: U.S. Government Printing Office, September 1986), p. 124.

[56] For instance, should prevention be measured by comparison with other comparable products, or by comparison with the same product in some previous base year? Using a base year as a standard of measurement may discriminate against companies that had already made significant reductions in product weight or toxicity before the base year, and reward those who did not.

[57] See Jesse H. Ausubel, "Industrial Ecology: Reflections on a Colloquium," *Proceedings of the National Academy of Sciences,* vol. 89, February 1992, p. 879; and Faye Duchin, "Industrial Input-Output Analysis: Implications for Industrial Ecology," p. 851 in the same volume.

[58] One example is the "Pellston-type" workshop organized by the Society of Environmental Toxicology and Chemistry in August 1990 to develop guidelines for life-cycle assessment methodology. See Society of Environmental Toxicology and Chemistry, "A Technical Framework for Life-Cycle Assessment," a workshop report, Washington, DC, January 1991.

[59] A review of the role of environmental factors in traditional cost accounting systems is provided by Rebecca Todd, "Accounting for the Environment: Zero-Loss Environmental Accounting Systems," presented at the National Academy of Engineering's Workshop on Industrial Ecology/Design for the Environment, Woods Hole, MA, July 13-17, 1992.

both industrial and engineering design). These case studies could provide excellent course materials for business schools and design schools.

Policy Research

Although many policy options have been suggested to control the environmental impacts of products, little is known about the costs and benefits of these options—especially the costs of monitoring and enforcement. In Germany, for example, requirements for manufacturers to take responsibility for recovering and recycling their packages appear to have been passed with little cost-benefit analysis.[60] Yet the costs may vary greatly depending upon the kind of product, its distribution network, and the waste management infrastructure. **Congress could require that EPA identify some of the more promising proposals being discussed around the world and analyze their likely costs and benefits.**

As indicated at the beginning of this chapter, the existing regime of environmental regulations, Federal and State procurement policies, and military specifications already have a profound influence on design decisions. **A useful starting point might be for Congress to direct EPA to coordinate a comprehensive review of how existing Federal and State regulations affect materials management decisions in the United States.** This could help consolidate reviews already ongoing in various agencies and provide a basis for government-wide administrative changes.

As shown in table 6-2, many States have initiated innovative programs to control the health and environmental impacts of products, including taxes on hard-to-dispose products, labeling requirements, and outright bans (e.g., Maine's ban on aseptic beverage containers). **As these State programs develop, Congress could direct EPA to evaluate their results. Congress may also wish to have EPA investigate the extent to which compliance with the growing diversity of State environmental** initiatives is imposing a serious financial burden on industry, with a view toward identifying areas where national standards are desirable.

CREDIBLE INFORMATION FOR CONSUMERS

The second unique role that the Federal Government can play in supporting green design is to ensure that consumers have reliable information about the environmental attributes of products, and to ensure that its own procurement of goods and services is consistent with environmental concerns. A significant fraction of consumers prefer to buy environmentally sound products,[61] and manufacturers are responding by touting the environmental benefits of their products, using terms like "recyclable," "biodegradable," and "ozone-safe."[62][63] But because the impacts of products on the environment are complex and multidimensional, there is tremendous potential for consumers to be confused by these diverse environmental claims.[64] In principle, LCA techniques can be used to determine the overall environmental quality of a product, but these techniques are still at an early stage of development, and it seems unlikely that definitive LCA results will be available for most products in the foreseeable future.

In general, two kinds of customers can be distinguished: individual consumers, and large-volume buyers for commercial firms, institutions, or government agencies. Options to address the information needs of these two groups are discussed below.

Individual Consumers

Environmental Advertising Claims

There is now a broad consensus on the part of industry, States, and environmental groups that Federal standards or guidelines of some sort are needed to regulate environmental advertising.[65] Industry wants national guidelines that prevent

---

[60] See "Environmentalism Runs Riot," *The Economist*, Aug. 8, 1992, p. 11.

[61] According to a survey conducted in 1990 by the Roper Organization for S.C. Johnson and Son (op. cit., footnote 14), 29 percent of participants reported purchasing a product because advertising or labeling said the product was environmentally safe.

[62] "Selling Green," *Consumer Reports*, op. cit., footnote 17.

[63] According to one survey, 26 percent of the 12,000 new household items launched in 1990 made some environmental claim. See Jaclyn Fierman, "The Big Muddle in Green Marketing," *Fortune*, June 3, 1991, p. 91.

[64] A survey by Environmental Research Associates of Princeton, NJ found that 47 percent of consumers dismiss environmental claims as "mere gimmickry." Jaclyn Fierman, ibid.

[65] For a good overview of the issues, see Ciannat M. Howett, "The 'Green Labeling' Phenomenon: Problems and Trends in the Regulation of Environmental Marketing Claims," *Virginia Environmental Law Journal*, vol. 11, spring 1992, p. 401.

deceptive claims and bring more uniformity to the current patchwork of State environmental labeling regulations.[66] However, industry generally opposes any regulations that would go beyond requirements for factual and verifiable statements about environmental attributes. Environmental groups want to establish a "floor" of Federal standards for advertisers' use of environmental terms that can be exceeded by States desiring to impose higher standards.[67] The underlying debate is between those who would treat environmental claims in the same way as any other form of advertising, and those who see it as a public policy tool for changing the behavior of manufacturers and consumers.[68]

In May 1991, a task force of 11 State attorneys general called for Federal standards for environmental advertising and recommended interim guidelines for use by manufacturers.[69] The task force recommended that environmental claims be as specific as possible, substantive (not trivial or irrelevant), and reflect current waste management options.

Also in May 1991, the Federal Trade Commission (FTC) published proposed guidelines for environmental advertising and held hearings in July to receive public comment. EPA published proposed guidelines for use of the terms "recyclable" and "recycled" and the use of the recycling emblem,[70] which industry critics charge go well beyond preventing deception. The FTC also joined with EPA and the U.S. Office of Consumer Affairs to form the Federal Interagency Task Force on Environmental Labeling, to coordinate Federal efforts.

In July 1992, the FTC issued final guidelines for environmental marketing claims.[71] The guidelines are intended to prevent deceptive environmental advertising, and are based on the principles that

claims of environmental benefits must be factual and verifiable. The guidelines are not intended to preempt State regulation of environmental advertising claims.

These FTC guidelines are an important step, especially because they encourage manufacturers to qualify broad claims of environmental benefits so as to be specific and verifiable. But Congress may wish to go beyond preventing deception; even if a claim is not overtly deceptive, it still may not convey sufficient information to enable consumers to evaluate the environmental benefits of the product or package. One criticism of FTC guidelines is that they do not provide standard definitions of environmental terms based on scientific criteria.[72] Critics argue that in any case, the FTC does not have the scientific expertise to evaluate such claims, and that the technical expertise of EPA is needed to develop credible scientific definitions.[73] **Congress could require that EPA work with the FTC to develop "official" definitions of environmental terms based on the best scientific information available.** In OTA's view, it is especially important to decide how terms relating to waste prevention (e.g., "source-reduced") should be defined and communicated to consumers.

Another criticism of the FTC guidelines is that they do not challenge manufacturers to make continuous improvements in order to be able to claim environmental benefits. **Congress could require EPA to develop minimum standards for unrestricted use of environmental terms in advertising.** For example, a product labeled as "recycled" might have to contain at least 10 percent post-consumer material.[74] These minimum standards could then be ratcheted up over time.

---

[66] In the absence of Federal standards, some 13 States have developed their own regulations on the use of environmental terms.

[67] See, e.g., testimony of the Environmental Defense Fund on environmental labeling and S. 615 before the Subcommittee on Environmental Protection of the Senate Committee on Environment and Public Works, July 31, 1991.

[68] Ciannet M. Howett, op. cit., footnote 65, p. 460.

[69] "The Green Report II: Recommendations for Responsible Environmental Advertising," May 1991.

[70] Environmental Protection Agency, "Guidance for the Use of the Terms 'Recycled' and 'Recyclable' and the Recycling Emblem in Environmental Marketing Claims," *Federal Register*, vol. 56, No. 191, Oct. 2, 1991, p. 49992.

[71] Federal Trade Commission, "Guides for the Use of Environmental Marketing Claims," July 1992.

[72] For example, a term like "degradable" might be defined differently by different manufacturers, giving consumers little information about the rate of degradation or the nature of the end products.

[73] Environmental Defense Fund, op. cit., footnote 67.

[74] This is the definition of "recycled" that has been adopted in California. EPA is seeking comment on a similar proposal. See EPA Guidelines, op. cit., footnote 70.

Industry generally objects to the idea of minimum standards, arguing that they could actually result in *less* information for consumers. For instance, if a product contains recycled content but does not quite meet the standard, manufacturers might be prevented from communicating this information to consumers. Environmental groups counter that this problem can be avoided if standards are applied only to *unqualified* use of terms; attributes not meeting minimum standards would have to be described in specific detail on the label (e.g., "This product contains 5 percent recycled industrial scrap and 5 percent post-consumer material.").

Eco-Labels

Even if consumers fully understand the environmental claims made by manufacturers, they are still faced with the problem of how to trade off one environmental benefit or cost versus another. For instance, it may be impossible for consumers to decide whether a product that contains "20 percent recycled content" is better for the environment than a similar one that uses "10 percent less packaging." Ideally, green products would carry a single indicator of overall environmental quality.[75]

According to the Organization for Economic Cooperation and Development, at least 22 countries are expected to develop green product labeling schemes by 1993.[76] As discussed in chapter 5, Germany, Canada, and Japan award an "eco-label" to products that are judged to have reduced environmental impact compared with competing products.[77] Properly constructed, environmental labels can provide consumers with an indicator of a product's overall environmental quality. Most analysts now agree that a properly constructed labeling program should be based on a life-cycle perspective (see chapter 4), rather than on a single environmental attribute such as recycled content. Initial efforts have focused on collecting an inventory of resource inputs and waste outputs. These inventories can provide useful insights, but the scope and interpretation of inventories completed so far have been controversial.[78] Nevertheless, at least the qualitative perspective of evaluating the entire life cycle of a product seems essential.

Although EPA once proposed the establishment of a U.S. national eco-label, it has since dropped the idea.[79] Instead, it is supporting research to develop LCA methods (see table 6-5). Two private sector labeling efforts, Green Seal and Scientific Certification Systems, are underway in the United States.[80] These efforts are still quite small, and at the present rate at which labeling guidelines are being developed, consumers should not expect to see eco-labels on a wide range of products in the near future.[81]

Congress could appoint a blue-ribbon commission to oversee the establishment of an independent, national eco-labeling program. A well-funded national program could accelerate the delivery of environmental information to consumers, especially if it borrowed the best from the experiences of existing programs around the world. A single national program would also have the credibility of the Federal Government behind it. **As an alternative, Congress could require EPA to develop standards for the certification of the product evaluation methods used by private eco-labeling programs.** A certification process would avoid the expense and bureaucracy of a national eco-labeling program, and avoid preempting private efforts that are already underway. For example, eco-label programs that are based on a legitimate life-cycle approach might receive government certification, while those based on a single environmental attribute might not.

By themselves, eco-labels are not likely to have a large impact on environmental quality. Indeed, only a small fraction of all products are likely to be

---

[75] A 1990 survey of 1,514 consumers conducted by *Advertising Age* and the Gallup organization found that 34 percent indicated that an eco-label program would have a great impact on their purchasing decisions. Cited in Ciannat M. Howatt, op. cit., footnote 65, p. 451.

[76] Catherine Arnst, "Some 22 Nationals Could Have 'Green Label' Schemes by '93," *Toronto Star*, Nov. 6, 1991, p. D6.

[77] In most countries, the label is awarded by a nongovernmental, independent institute according to strict national rules.

[78] *Consumer Reports*, op. cit., footnote 17.

[79] Hannah Holmes, "The Green Police: In the Environmental Holy War, Who Can Tell the Good Guys From the Bad Guys?" *Garbage*, September-October 1991, p. 44.

[80] Amy Lynn Salzhauer, "Obstacles and Opportunities for a Consumer Ecolabel," *Environment*, vol. 33, No. 9, November 1991, p. 10.

[81] Ibid., p. 36.

considered for an eco-label.[82] There is also a question about the extent to which manufacturers of environmentally harmful products will be motivated to redesign products and processes to become eligible for a label. Nevertheless, eco-labels may provide public policy benefits that reach beyond the labeled products themselves. A highly visible eco-labeling program could become a useful educational tool to raise consumer awareness about the environment that could spill over to other purchasing decisions.

Institutional, Commercial, and Government Buyers

OTA estimates that over 40 percent of all goods and services (by value) produced in the U.S. economy are "intermediate" goods and services (e.g., industrial equipment, chemicals, etc.) that are purchased by businesses, institutions, or government agencies, rather than by individual consumers.[83] Therefore, these large-volume buyers are an important target for environmental information.

While the "green preferences" of individual consumers have been the subject of numerous studies, OTA is unaware of any systematic studies on how environmental concerns are factored into the purchasing decisions of commercial or institutional buyers. This is an important area for further research (see section on institutional and behavioral research above).

There is anecdotal evidence that these large-volume buyers are beginning to request more information about the environmental attributes of products and packaging. But since these intermediate goods are not advertised in the same way as consumer goods, eco-labels or environmental advertising standards may not be appropriate. In the case of chemicals, EPA has studied the possibility of increasing information to users of TRI chemicals by

expanding Material Safety Data Sheets to include environmental hazards,[84] or requiring manufacturers to provide "product stewardship" information to their customers.[85] In Europe, the idea of requiring manufacturers to provide a "product environmental profile" to their customers is being explored (see chapter 5).[86]

These proposals should be evaluated carefully and full advantage should be taken of voluntary industry efforts that are already ongoing.[87] Federal regulations requiring the transmission of additional environmental information between suppliers and manufacturers could create additional paperwork without addressing the specific concerns of individual buyers. However, it is important that the Federal Government incorporate environmental criteria into its own purchasing decisions.

Government Procurement

About 20 percent of the purchases of all U.S. goods and services is made by government at the Federal, State, and local levels. Section 6002 of RCRA requires EPA to establish procurement guidelines for government agencies to purchase products made with recovered materials. At this writing, EPA had published guidelines for paper products, lubricating oils, retreaded tires, building insulation, and cement or concrete containing fly ash. Several more guidelines are expected in 1992.[88] **Congress could require EPA to identify additional product categories and establish deadlines for issuance of "green" procurement guidelines.**

To date, government procurement guidelines for green products have been based almost exclusively on recycled content. In the future, it will be important to broaden these guidelines to include other environmental attributes, especially waste prevention (toxicity reduction, energy efficiency, etc.). **Congress could require that EPA undertake**

[82] In 1991, after 13 years in operation, Germany's Blue Angel program had awarded eco-labels to some 3,600 products in 66 product categories. However, more than half of the labels awarded fall into only four product categories.

[83] This estimate was obtained from the Department of Commerce's Use of Commodities Table of 1987. Dividing total intermediate use by total commodity output yields a ratio of 43.6 percent.

[84] Required by the Occupational Safety and Health Administration.

[85] David Hanson, "EPA Develops Product Stewardship, Hazard Communication Regulations," *Chemical and Engineering News*, Nov. 19, 1990.

[86] A product profile is a qualitative description of the life-cycle environmental impacts of a product, intended for use by professional buyers, rather than individual consumers.

[87] See, e.g., Janice R. Long, "Standard for Material Safety Data Sheets in the Offing," *Chemical and Engineering News*, May 18, 1992, p. 7.

[88] Testimony of Richard D. Morgenstern, Acting Assistant Administrator, Office of Policy Planning and Evaluation, EPA, before the Subcommittee on Oversight of Government Management, Senate Committee on Governmental Affairs, Nov. 8, 1991.

studies to determine how procurement guidelines might be broadened to account for waste prevention.

In response to congressional pressure, the General Services Administration (GSA) has begun to highlight the environmental attributes of products in its regular supply catalogs.[89] This has helped to ensure that procurement agents in various agencies have access to environmental information on the products they buy. **Congress could formalize this process by requiring that all Federal procurement catalogs contain information on environmental attributes alongside performance and cost information.**

MARKET DISTORTIONS AND ENVIRONMENTAL EXTERNALITIES

The third major area in which Congress can encourage green design is by shaping environmental policies that better account for the environmental impacts of products throughout their life cycle. Providing better information to designers and consumers on the environmental impacts of materials and processes is important, but if this information is not backed up by appropriate price signals, environmental concerns are likely to be overwhelmed by many other design requirements and consumer demands.

Economists have long argued that efficient use of energy and resources requires that the prices of goods and services reflect their true social (and environmental) costs.[90][91] These costs are partially accounted for through health and environmental laws such as those in table 6-1. For example, emissions control technologies required by the Clean Air Act raise the price of electric power and automobiles. Nationwide, it is estimated that compliance with pollution control laws costs industry and consumers $115 billion per year.[92]

But most observers agree that many environmental costs remain external to economic transactions, and in some cases government policies distort market price signals. On the production side, there are government subsidies or special tax treatment for the extraction of virgin materials (e.g., below-cost timber sales and mineral depletion allowances);[93] and many "non-hazardous" industrial solid wastes (e.g., mine tailings or manufacturing wastes that are managed on-site) with significant environmental impacts are not regulated at the Federal level.[94] On the consumption side, consumers often do not pay the full environmental costs of products that are consumed or dissipated during use (e.g., fuels, cleaners, agricultural chemicals),[95] or the full cost of solid waste disposal.[96]

There are two general policy mechanisms for internalizing environmental costs: regulations and economic instruments. Historically, the basis of environmental policy in the United States has been regulation (see, for example, table 6-1). But in recent years, there has been a growing interest in the use of market-based incentives such as pollution taxes, tradable pollution permits, and deposit-refund systems, that can—in principle at least—provide the same environmental protection as regulations at less cost.[97] Table 6-6 presents a menu of regulatory and market-based incentives that have been proposed to internalize the environmental costs associated with the flow of goods and materials through the economy. These options are organized according to their point of greatest impact on the materials life cycle. Each could have an impact on product design, but an

---

[89] See, e.g., GSA Supply Catalog, May 1992. See also "Buying Green: Federal Purchasing Practices and the Environment," a hearing of the Subcommittee on Oversight of Government Management of the Senate Committee on Governmental Affairs, Nov. 8, 1991.

[90] For a recent review, see William D. Nordhaus, "The Ecology of Markets," *Proceedings of the National Academy of Science*, vol. 89, February 1992, p. 843.

[91] For a highly readable discussion of environmental policy instruments to protect the environment from an economist's perspective, see Frances Cairncross, *Costing the Earth: The Challenge for Governments, the Opportunities for Business* (Boston, MA: Harvard Business School Press, 1992).

[92] Environmental Protection Agency, "Environmental Investments: The Costs of a Clean Environment," EPA 230-12-90-084, December 1990.

[93] Jessica Matthews, "Oh, Give Me a Home Where the Subsidies Roam," *Washington Post*, Oct. 3, 1991.

[94] U.S. Congress, Office of Technology Assessment, *Managing Industrial Solid Wastes From Manufacturing, Mining, Oil and Gas Production, and Utility Coal Combustion—Background Paper*, OTA-BP-O-82 (Washington, DC: U.S. Government Printing Office, February 1992).

[95] See, e.g., Harold M. Hubbard, "The Real Cost of Energy," *Scientific American*, vol. 264, No. 4, April 1991, p. 36.

[96] A. Clark Wiseman, "Impediments to Economically Efficient Solid Waste Management," *Resources*, fall 1991, p. 9.

[97] Robert W. Hahn and Robert N. Stavins, "Incentive-Based Environmental Regulation: A New Era From an Old Idea?" Energy and Environmental Policy Center Discussion Paper, John F. Kennedy School of Government, Harvard University, August 1990.

analysis of the design implications of all of them is beyond the scope of this report. Here we focus especially on those options that would affect product design directly: i.e., options primarily affecting the manufacturing stage of the life cycle. Many of these instruments have been discussed in detail elsewhere.[98]

Recycled Content

With the proliferation of State and local recycling collection programs in recent years, cities are collecting recyclables at a pace that far exceeds the use of these recovered materials in new products. Cities are now faced with increasing costs of managing recovered materials at a time when they are already strapped financially.[99] As more and more large cities implement collection programs, the volume of recovered materials can be expected to increase; without new markets for these materials, prices will drop. This has led to increasing pressure on Congress to enact legislation to create markets for recovered materials through recycled content requirements.

Recycled content requirements can help to solve the immediate problem of the lack of markets for recyclables. But by creating markets for recovered materials through regulation, policymakers are imposing a predetermined solution to the solid waste problem that ignores market forces.[100] This solution may be inefficient for several reasons:

- With many thousands of products likely to be covered by such regulations, the transaction costs of administration, monitoring, and enforcement on a per-product basis may be unacceptably high.

- Across-the-board content requirements do not account for the fact that some companies may be able to incorporate recycled content more cheaply than others, or that costs may vary significantly by geographical region.
- Finally, by focusing exclusively on a single environmental attribute, recycled content requirements may preclude environmentally preferred designs, especially those featuring waste prevention.

Congress could choose not to address this problem, in which case many communities may be forced to curtail their recycling collection programs until stronger markets for these materials develop. If Congress does choose to pursue recycled content requirements, either in government procurement programs or as part of RCRA, it can address the inefficiencies noted above in several ways.

Crediting Waste Prevention

Congress can exempt products from recycled content requirements that feature waste prevention.[101] This would provide more flexibility to manufacturers, but the viability of this option depends on developing criteria for measuring waste prevention. For instance, how should the various aspects of waste prevention (e.g., weight or volume reduction, toxicity reduction, and energy efficiency) be factored in? And what should be the baseline for measuring reductions? A system of complicated exemptions for waste prevention could make recycled content regulations administratively unworkable, especially if applied on a per-product basis. **One alternative would be to offer companies the option of avoiding per-product regulations by committing to companywide reductions, and**

---

[98] See, e.g., Resource Conservation Committee, "Choices for Conservation," Final Report to the President and Congress, SW-779, July 1979 (available from the U.S. EPA Engineering Research Center Library, Cincinnati, OH); Robert N. Stavins, Project Director, "Project 88, Harnessing Market Forces To Protect Our Environment: Initiatives for the New President," Washington, DC, December 1988; Robert N. Stavins, Project Director, "Project 88—Round II, Incentives for Action: Designing Market-Based Environmental Strategies," Washington, DC, May 1991; Environmental Protection Agency Science Advisory Board, "Reducing Risk: Setting Priorities and Strategies for Environmental Protection," (especially Appendix C), EPA-SAB-EC-90-021C, September 1990, U.S. Environmental Protection Agency, "Economic Incentives: Options for Environmental Protection," PM-220, Washington, DC, March 1991; Organisation for Economic Cooperation and Development, "Environmental Policy: How To Apply Economic Instruments," Paris, 1991.

[99] The National Solid Waste Management Association noted in a recent survey of the nation's cities that solid waste management costs are second only to education in public expenditure of funds. David Ruller, Recycling Coordinator for the City of Alexandria, VA, personal communicaton, August 1992.

[100] "How To Throw Things Away," *The Economist,* Apr. 13, 1991, p. 17.

[101] The Massachusetts Public Interest Research Group (MASSPIRG) has developed legislation proposing that packaging be considered "green" if it met one of several alternative criteria: made from a specified percentage of recovered material, made from a material that was recycled at a specified rate, or reduced in weight or volume by a specified percentage. Such "MASSPIRG bills" have been introduced in several States, and the idea became part of the RCRA reauthorization debate in the 102d Congress.

Table 6-6—Policy Options That Could Affect Materials Flows

Life-cycle stage	Regulatory instruments	Economic instruments
Raw material extraction and processing	Regulate mining, oil, and gas non-hazardous solid wastes under the Resource Conservation and Recovery Act (RCRA). Establish depletion quotas on extraction and import of virgin materials.	Eliminate special tax treatment for extraction of virgin materials, and subsidies for agriculture. Tax the production of virgin materials.
Manufacturing	Tighten regulations under Clean Air Act, Clean Water Act, and RCRA. Regulate non-hazardous industrial waste under RCRA. Mandate disclosure of toxic materials use. Raise Corporate Average Fuel Economy Standards for automobiles. Mandate recycled content in products. Mandate manufacturer take-back and recycling of products. Regulate product composition, e.g., volatile organic compounds or heavy metals. Establish requirements for product reuse, recyclability, or biodegradability. Ban or phase out hazardous chemicals. Mandate toxic use reduction.	Tax industrial emissions, effluents, and hazardous wastes. Establish tradable emissions permits. Tax the carbon content of fuels. Establish tradable recycling credits. Tax the use of virgin toxic materials. Create tax credits for use of recycled materials. Establish a grant fund for clean technology research.
Purchase, use, and disposal	Mandate consumer separation of materials for recycling.	Establish weight/volume-based waste disposal fees. Tax hazardous or hard-to-dispose products. Establish a deposit-refund system for packaging or hazardous products. Establish a fee/rebate system based on a product's energy efficiency. Tax gasoline.
Waste management	Tighten regulation of waste management facilities under RCRA. Ban disposal of hazardous products in landfills and incinerators. Mandate recycling diversion rates for various materials. Exempt recyclers of hazardous wastes from RCRA Subtitle C. Establish a moratorium on construction of new landfills and incinerators.	Tax emissions or effluents from waste management facilities. Establish surcharges on wastes delivered to landfills or incinerators.

SOURCE: Office of Technology Assessment, 1992.

requiring them to document compliance with these agreements in publicly available databases.[102]

Tradable Recycling Credits

Another option for reducing the burden of industry compliance with recycled content requirements would be to couple them with a tradable recycling credit mechanism similar to the emissions trading program for sulfur dioxide under the Clean Air Act Amendments of 1990.[103] Manufacturers would be required either to use a specified percentage of recycled content in their products, or to purchase recycling credits from other manufacturers who exceed the percentage requirement.

Such tradable recycling credit mechanisms encourage those manufacturers that can incorporate recovered materials most cheaply to do so. However, due to the administrative costs of setting up and monitoring these programs, they are not feasible for all of the many thousands of products on the market.[104] Rather, they may be most suitable for a limited number of materials or waste streams of special concern, e.g., old newspapers, used oil, or

[102] Model legislation developed by the Source Reduction Task Force of the Coalition of Northeastern Governors (CONEG) would allow companies to avoid all packaging recycling requirements by committing to reduce packaging on a companywide basis by 15 percent between 1988 and 1996. See CONEG Source Reduction Task Force—Model Legislation, "An Act Concerning Reduction in Packaging Waste," Feb. 11, 1992.

[103] See "Project 88—Round II," op. cit., footnote 98, p. 55.

[104] For a discussion of the appropriate use of marketable permits, see Organisation for Economic Cooperation and Development, op. cit., footnote 98.

automobile batteries.[105] **Congress could mandate that EPA set up a limited number of pilot tradable recycling credit programs to evaluate the effectiveness of this approach.**

A Market-Based Alternative

The extraction of raw materials and their initial processing are two of the most environmentally destructive phases of a product's life cycle.[106] Yet many of these environmental costs are not reflected in the price of materials. **As an alternative to recycled content requirements, Congress may wish to move toward a system of indirect incentives aimed at internalizing the environmental costs of virgin materials use, thus making virgin materials more expensive and the use of recovered materials more economically attractive.**[107] This might include eliminating subsidies and special tax treatment for the extraction of virgin materials, taxing the production of virgin materials of special concern, or regulating more strictly the wastes and other environmental impacts of extractive industries.[108]

Such a market-based strategy has the advantage that it does not impose a predetermined solution on the solid waste problem, and would begin to internalize the costs of some of the most environmentally destructive practices. However, the size of its impact on the solid waste problem and the timing of that impact is much less predictable. Several studies suggest that simply removing government subsidies for virgin materials is unlikely to change the price of processed materials by more than a few percent.[109] [110] Attempting to replicate the incentives of recycled content requirements through taxes on virgin materials might require taxes to be so high as to cause significant economic disruption in the

domestic materials extraction industries, with serious implications for U.S. resource security. Nevertheless, given that current policies were established at a time when the goal was to encourage the exploitation of resources,[111] it is appropriate for Congress to reevaluate these policies in light of current concerns about the environmental impacts of resource use and ecological sustainability.

Use of Hazardous Chemicals

Since the 1940s, when the chemical industry began an era of explosive growth, more than 60,000 chemical substances have been synthesized, and more than 1,000 new chemicals are proposed for manufacture each year.[112] These chemicals are responsible in large part for the high standard of living in industrialized countries, and for many of the conveniences of modern life. Contemporary food production, medicines, building materials, and many consumer products (e.g., nylon hosiery and laundry detergents) depend on use of these chemicals.

This dramatic growth in chemical use has also raised health and environmental concerns. For the most part, these chemicals pass through the economy quickly, whether in the form of industrial wastes or products.[113] Some have very long lifetimes in the environment (e.g., CFCs and polychlorinated biphenyls (PCBs)) and may become distributed globally.[114] In some cases, a hazardous substance may achieve widespread use before its health or environmental implications are realized; for example, CFCs were believed to be quite safe at the time they were introduced. Toxic substances initially released in low concentrations may also become reconcentrated in sediments or through bioaccumu-

---

[105] Legislation to establish tradable recycling credit programs for old newspapers, tires, used oil, and automobile batteries was introduced in the 102d Congress by Representative Esteban Torres. See, e.g., H.R. 872.

[106] John E. Young, ''Tossing the Throwaway Habit,'' *World Watch*, May-June 1991, p. 26.

[107] Steven Kraten, ''Market Failure and the Economics of Recycling,'' *Environmental Decisions*, April 1990, p. 20.

[108] The advantages and disadvantages of these options are discussed extensively in the references of footnote 98.

[109] For a brief discussion of the impact of virgin material subsidies on recycling, see U.S. Congress, Office of Technology Assessment, *Facing America's Trash: What Next for Municipal Solid Waste*, OTA-O-424 (Washington, DC: U.S. Government Printing Office, October 1989), p. 200.

[110] Other observers counter, though, that the largest government subsidies go to the energy and transportation sectors, not to virgin materials per se.

[111] For example, the law that governs the extraction of gold, silver, and other ''hard rock'' minerals is the General Mining Law of 1872. Bills to reform the Mining Law were introduced in both the House and Senate in the 102d Congress.

[112] Michael Shapiro, ''Toxic Substances Policy,'' in *Public Policies for Environmental Protection*, Paul R. Portney (ed.), op. cit., footnote 2, p. 195.

[113] Robert U. Ayres, ''Industrial Metabolism,'' *Technology and Environment* (Washington DC: National Academy Press, 1989), p. 23.

[114] Curtis C. Travis and Sheri T. Hester, ''Global Chemical Pollution,'' *Environmental Science and Technology*, vol. 25, No. 5, 1991, p. 814.

lation to levels that pose significant risks to human health.[115]

In 1976, Congress passed the Toxic Substances Control Act (TSCA) to address these concerns. Yet little is known about the long-term implications of the dissipative use of these substances for human health and the environment. Toxicity data are lacking on many, if not most of the chemical products used in the United States.[116] In 1991, GAO reported that 15 years after the enactment of TSCA, EPA had received test results for only 22 chemicals, and had assessed the results for only 13 of the 22.[117]

Manufacturers have begun to respond to these health and environmental concerns in a variety of ways, such as the ''Responsible Care'' program of the Chemical Manufacturers Association (see table 6-3). As of February 1992, 734 companies had joined EPA's 33/50 Program, pledging to reduce their releases of 17 toxic materials by 50 percent (relative to 1988) by 1995.[118] But in spite of these efforts, large volumes of hazardous chemicals continue to flow through the economy into the environment. According to data collected in 1990 on industrial use of hazardous substances in the State of New Jersey, for example, at least 83 percent of the cadmium, 92 percent of the nickel, and 99 percent of the mercury used by industry was converted into products (e.g., paints, coatings, plastics, and batteries), not released as wastes.[119] These heavy metals are released to the environment when these products are discarded; however, these environmental releases are not addressed by programs such as the 33/50 program, which is concerned only with industrial waste streams. This example illustrates that if we are concerned about the dissipation of hazardous materials into the environment, we must be concerned not only with industrial wastes, but with the use of these materials in products as well.

Toxics Use Reduction

Recognizing the importance of toxic materials flows in products as well as in industrial wastes, environmental groups are promoting reduction in industry's use of toxic chemicals in the first place.[120] The rationale is that once toxic materials are introduced into the economy, they are likely to be released into the environment. Therefore, environmental groups argue, the best way to prevent toxic releases is to limit the use of these materials from the outset. Some advocates envision a world in which certain toxic materials would be ''sunsetted'' or phased out entirely.[121]

The distinction between waste minimization and toxics use reduction is important because toxics use reduction is a much more radical concept than waste minimization (box 6-B). Whereas Federal policy has long been concerned with protecting the environment from the release of hazardous and nonhazardous wastes by industrial generators (e.g., EPA's 33/50 Program), the choice of what materials should be used in products has usually been a private sector decision. Thus, toxics use reduction implies government intrusion into areas that have traditionally been considered the province of private industry. In other words, toxics use reduction involves a more prescriptive approach to product design than does waste minimization.

Policy Approaches

The use of hazardous or toxic chemicals must be understood in the context of risks and benefits.[122] Clearly, the environmental risks of using some materials are so great that they outweigh any possible benefits, and they must be banned—as in

[115] For example, mercury volatilized by fossil fuel burning or municipal solid waste incineration remains in the atmosphere for about a year. After mercury from the atmosphere is deposited in lakes, it is methylated and bioaccumulates in fish as methyl mercury, the form most toxic to humans. About 15 percent of Michigan lakes, 30 percent of Wisconsin lakes, and 50 percent of Florida lakes contain fish with mercury levels exceeding State health standards. Curtis C. Travis and Sheri T. Hester, op. cit., footnote 114, p. 816.

[116] Michael Shapiro, op. cit., footnote 112, p. 221.

[117] U.S. Congress, General Accounting Office, *Toxic Substances: EPA's Chemical Testing Program Has Not Resolved Safety Concerns*, GAO/RCED-91-136 (Washington, DC: U.S. Government Printing Office, 1991), p. 2.

[118] Environmental Protection Agency, ''33/50 Program Pledges on the Rise,'' *Pollution Prevention News*, March/April 1992, p. 1.

[119] Data supplied by Andrew Opperman, New Jersey Department of Environmental Protection and Energy, personal communication, August 1992.

[120] See, e.g., testimony of Hillel Gray, National Environmental Law Center, before the Subcommittee on Transportation and Hazardous Materials of the House Committee on Energy and Commerce, Mar. 10, 1992.

[121] Jeffrey A. Foran, ''The Sunset Chemicals Proposal,'' *International Environmental Affairs*, vol. 2, No. 4, fall 1990, p. 303.

[122] This principle is at the root of the Toxic Substances Control Act and the Federal Insecticide, Fungicide, and Rodenticide Act, which regulate chemicals and pesticides on the basis of ''unreasonable risk'' to health or the environment.

Box 6-B—*Toxics Use Reduction vs. Waste Minimization*

Waste minimization and toxic use reduction are related concepts, but they are not identical. Waste minimization refers to all activities that reduce the quantity or toxicity of waste released from a facility to the environment. As such it is concerned with reducing the waste *outputs* of industrial processes. Toxic use reduction refers to reducing the *inputs* of toxic materials into industrial processes, thereby avoiding their release as wastes *or in products*.

Industry has been generally supportive of the idea of waste minimization, at least in principle. Companies object strongly, though, to government requirements for toxic use reduction. They argue that society's legitimate concern is with the release of toxic materials, not their use *per se*. For example, two toxic chemicals can react to produce a nontoxic product; and toxic solvents can be recovered and reused many times. Thus, they argue, the mere use of a toxic material may not affect the environment. Furthermore, companies argue that the term "toxic" is imprecise because most substances are toxic in sufficient concentrations, while some highly toxic chemicals can be beneficial in low concentrations. Finally, industry argues that regulations restricting the use of materials would hurt their international competitiveness, since the same restrictions would not apply to their competitors overseas.

Environmentalists counter that industry cannot be entrusted with protecting the environment from toxic materials, especially when it is not profitable to do so. They point to historical examples of polluted rivers and abandoned toxic waste dumps. And even if toxic materials are released in small quantities, they may persist for a long time in the environment and become reconcentrated in sediments or through bioaccumulation. Little is known about the risks of long-term exposure to low concentrations of toxic chemicals.

These two views—waste minimization and toxic use reduction—illustrate the clash between two of the philosophical paradigms discussed in chapter 3. Waste minimization, with its concern with industrial waste outputs, arises from the environmental protection paradigm. Toxic use reduction, on the other hand, with its precautionary emphasis on resource inputs, reflects the eco-development paradigm.

SOURCE: Office of Technology Assessment, 1992.

---

the case of CFCs. For most chemical substances, though, more flexibility is appropriate. Products that use toxic materials can perform socially useful functions or even have (comparative) environmental benefits. For instance, the recently discovered high-temperature superconductors could potentially lead to more efficient power generation and transmission, resulting in less pollution from power plants. Yet the synthesis of these superconductors involves use of toxic chemicals, and the materials themselves contain a variety of toxic heavy metals; for instance, the compound currently with the highest critical transition temperature is based on thallium, a highly toxic heavy metal.[123]

On the other hand, it must be recognized that there is considerable uncertainty about the health and environmental impacts of the dissipative use of hazardous or toxic materials. As noted above, information on the toxicity and long-term health effects of most chemicals is sketchy at best, and the environmental risks to ecosystems have hardly been considered. These uncertainties suggest that a precautionary policy that encourages designers to avoid the dissipative use of hazardous materials (insofar as possible) is warranted.

More than a dozen States have enacted laws that promote toxics use reduction or related approaches.[124] The Massachusetts Toxics Use Reduction Act, which is widely agreed to be the most aggressive, requires industrial facilities to develop toxics use reduction plans and document progress toward self-set goals. The overall goal of the legislation is to reduce the use of listed toxic chemicals by 50 percent by 1997. To protect proprietary information, the plans themselves are confidential, although the plan summaries and the goals are to be made public.[125]

---

[123] U.S. Congress, Office of Technology Assessment, *High-Temperature Superconductivity in Perspective*, OTA-E-440 (Washington, DC: U.S. Government Printing Office, April 1990).

[124] William Ryan and Richard Schrader, "An Ounce of Toxic Pollution Prevention: Rating States' Toxic Use Reduction Laws," available from the Center for Policy Alternatives, Washington, DC, Jan. 17, 1991, p. 1.

[125] Ken Geiser, "The Greening of Industry," *Technology Review*, August/September 1991, p. 68.

California has tried another approach aimed at informing consumers of the use of toxic chemicals in products. Under the Safe Drinking Water and Toxic Enforcement Act of 1986 (Proposition 65), products that contain even minute amounts of any of 420 chemicals determined to be carcinogenic or posing reproductive toxicity must be labeled with warnings.[126] This has stimulated some companies to reformulate products to avoid the labeling requirements.[127] However, because so many chemicals are covered, and in such low concentrations, the effect of the labeling requirements may be to desensitize consumers to actual risks.[128]

In addressing health and environmental concerns relating to toxic or hazardous chemical use in products, Congress can choose a variety of options, ranging from further research to mandatory toxics use reduction requirements. **By initiating a research program to identify high-risk materials and products, and to model the flows of these materials through the economy (see research discussion above), Congress can ensure that regulations result in cost-effective risk reduction.**

Congress can act to increase available information about the flows of toxic materials by expanding industry's reporting requirements under TRI to include additional facilities, industrial sectors, and chemicals not covered under the original law. The 1986 Superfund Amendments and Reauthorization Act requires manufacturers in 20 manufacturing industries to report annually to EPA on their releases of 322 chemicals or chemical categories. However, many environmental releases of these chemicals are not covered under TRI. Not included are nonmanu-

facturing sources such as mines, waste treatment plants, public utilities, farms, and government facilities. Manufacturers with fewer than 10 employees or using less than 10,000 pounds of TRI chemicals annually are exempted from reporting. Critics also charge that hundreds of chemicals listed as toxic under other environmental laws are excluded from the TRI reports.[129]

Congress could expand facilities' reporting requirements under TRI to include the *use* of toxic materials, not just their releases to the environment. This "materials accounting" approach could lead to a valuable database on toxic chemical flows.[130] However, comprehensive reporting on the use of hundreds of chemicals for thousands of facilities would involve a huge paperwork burden, both for companies and for EPA reviewers.[131] Unless these requirements are narrowly targeted on chemicals or materials of special concern, they would significantly increase industry's reporting costs, and might not result in a significant reduction of environmental risk.

Congress could mandate a national requirement for industry toxics use reduction plans, modeled on the Massachusetts Toxics Use Reduction Act. Again, however, the law would have to be carefully structured to make the paperwork burden manageable. **If Congress does decide to pursue mandatory toxics use reduction, it may wish to consider market-based incentives such as tradable toxic use permits to achieve reductions at the lowest cost to industry.**[132] This approach was used successfully in the phase-out of leaded gasoline

---

[126] William S. Pease, "Chemical Hazards and the Public's Right to Know: How Effective Is California's Proposition 65?" *Environment*, vol. 33, No. 10, December 1991, p. 12.

[127] For example, Gillette reformulated its Liquid Paper product to remove trichloroethylene, one of the listed chemicals. The substitute was 1,1,1 trichlorethane, which is not on the list. Robert Healey, Gillette, personal communication, August 1992.

[128] The law states that listed chemicals in foods must not be present at a level greater than one one-thousandth of the level at which there are no observable health effects. Industry representatives claim that this is too restrictive, arguing, for example, that ethyl alcohol, one of the listed chemicals, is naturally present in soft drinks, carrots, ice cream, and bread at levels that would trigger a warning under Proposition 65. See Conrad B. Mackerron, "Industry Is Learning To Live With Proposition 65," *Chemicalweek*, July 12, 1989, p. 19.

[129] According to one estimate, TRI excludes 140 chemicals regulated as hazardous under RCRA; 64 substances listed as hazardous under the Clean Air Act; 56 priority pollutants under the Clean Water and Safe Drinking Water Acts; 69 special review pesticides under FIFRA; and hundreds of probable carcinogens and reproductive toxicants listed by scientific authorities and government agencies. See Hillel Gray, op. cit., footnote 120.

[130] For example, New Jersey's Worker and Community Right to Know Act of 1983 requires industry to report inputs and outputs of 165 hazardous chemicals, all of which are on the TRI list. These materials accounting data are necessary to track the flows of these chemicals through the economy and into the environment, whether in the form of products or waste streams.

[131] The petroleum industry, for instance, argues that crude oil contains millions of different hydrocarbons and other naturally occurring compounds that are never fully separated during the manufacturing process. Petroleum products such as gasoline, fuel oil, and others are also complex mixtures that do not have standard compositions. Accounting for all of these chemicals would be impractical. See testimony of the American Petroleum Institute before the Subcommittee on Transportation and Hazardous Materials of the House Committee on Energy and Commerce, Mar. 10, 1992.

[132] Molly K. Macauley and Karen L. Palmer, "Incentive-based Approaches to Regulating Toxic Substances," *Resources*, summer 1992, p. 5.

during the early 1980s.[133] Finally, if new bans are deemed necessary, they can be targeted on specific products, rather than generic materials. This can help to target specific risks, while not foreclosing the economic and environmental benefits that these materials may have in other applications.

Product Taxes

Environmental product fees or taxes are—in principle, at least—an efficient way to encourage designers and consumers to make greener choices.[134] Taxes can be applied to hard-to-dispose products or to products that pose special risks as a result of their use. The best example in the United States is the excise tax on CFCs, which is intended to remove windfall profits as the production of these chemicals is phased down under a marketable permit system.[135] As discussed in chapter 5, several European countries impose hefty taxes on nonreturnable beverage containers and other packaging to encourage returnable and reusable packaging. Several countries in Europe impose a tax on leaded gasoline (with a corresponding subsidy for unleaded gasoline), which has resulted in a significant decline in demand for leaded gas.[136]

More often, fees are imposed on products to raise funds for recycling or safe disposal programs, and are too small to influence product design decisions. Several States have fees on products that cause special waste problems, such as tires, batteries, and used oil. At this writing, Florida is the only State with an advance disposal fee on packaging.[137]

Industry has lobbied heavily to quash proposals for new environmental product taxes, arguing that taxes on narrow categories of products (e.g., packag-ing) are unfair, while taxes on a large number of different products could involve unacceptable administrative costs. There is, however, widespread agreement among industry and environmentalists that weight- or volume-based trash disposal fees provide an excellent incentive for consumers to send less trash to the landfill—provided they have access to curbside recycling programs (for which no disposal fee is charged). An increasing number of communities have implemented these pay-per-can programs.[138] However, while these programs may encourage the separation of trash for recycling, they seem unlikely to influence consumer buying habits (and thus product design) in a dramatic way, because solid waste disposal costs are relatively small compared with the price of most products.[139]

In the short term, Congress could set up a national waste disposal fee that would fund a grant program for research, demonstration, and education projects for clean manufacturing technologies and green product design in universities, national laboratories, and industry. For example, a Federal charge of $1 per ton of municipal solid waste delivered to landfills and incinerators would raise on the order of $100 million annually. Such a charge would not discriminate against specific products, and the infrastructure for collecting the charge already exists in most states, so collection costs would likely not be prohibitive.[140]

In the longer term, if Congress decides to address energy conservation and global warming concerns through an environmental tax on fossil fuels, this could have a dramatic impact on product design, since fuels are consumed at every stage of the product life cycle. Such a tax could encourage not only the design of more energy-

---

[133] R. Hahn and G. Hester, "Marketable Permits: Lessons for Theory and Practice," *Ecology Law Quarterly*, vol. 16, No. 2, 1989.

[134] Terry Dinan, "Solid Waste: Incentives That Could Lighten the Load," *EPA Journal*, vol. 18, No. 2, May/June 1992, p. 12.

[135] David Lee, "Ozone Loss: Modern Tools for a Modern Problem," *EPA Journal*, May/June 1992, p. 16.

[136] See Environmental Resources Ltd., "Environmentally Sound Product Design: Policies and Practices in Western Europe and Japan," contractor report prepared for OTA, July 1991, p. 45.

[137] The fee is $0.01 per container by 1992 unless the container material reaches a 50 percent recycling rate.

[138] See, e.g., Environmental Protection Agency, "Economic Incentives: Options for Environmental Protection, op. cit., footnote 98, p. 2-7.

[139] Consider a family of four that annually purchases $10,000 of goods requiring disposal. Annual discards (at a rate of 4 pounds per person per day, the national average) amount to 2.92 tons of trash. At a weight-based fee of $100 per ton, this amounts to an annual trash bill of $292, about 3 percent of purchases. By careful shopping for recyclable and light-weight products and packaging, consumers might save 10 percent on their trash bill (about $30 per year), or 0.3 percent of total purchases.

[140] Denmark imposes a national tax on the weight of solid wastes delivered to landfills and incinerators. The tax is earmarked to pay for recycling and environmental research programs.

[141] As one example, building aircraft with new, light-weight composites can significantly improve their fuel efficiency, but with the decline of real jet fuel prices since the late 1970s, the higher initial cost of composites compared with aluminum cannot be recouped through fuel savings.

efficient products, but more material-efficient products as well.[141]

Manufacturer Take-Back

Take-back regulations give manufacturers responsibility for recovering and recycling the products they produce. By shifting the burden of solid waste management from beleaguered municipal governments to industry, the costs of solid waste management are internalized and manufacturers have direct incentives to design products that are recyclable. As discussed in chapter 5, Germany has established a take back program for packaging, and is considering the idea for a variety of durable goods as well. The takeback idea appears to be gathering momentum throughout Europe,[142] and many observers believe its introduction in the United States is just a matter of time.[143]

Manufacturer take-back regulations have considerable intuitive appeal. By assigning manufacturers the responsibility for recovering their own products, rather than telling them how to do it, manufacturers have some flexibility to find the least-cost solution. This may involve collecting and recycling the product themselves, or paying a third party to do so.

Durable goods may be especially good candidates for take-back programs, because they are inherently longer lasting, are generally made from higher value materials, and often consist of "knowledge-intensive" components that command a high recovery value. Indeed, some manufacturers of leased office equipment have already initiated design for recycling and remanufacturing programs (see chapter 3). Products that pose special solid waste disposal problems, such as batteries and tires, may also be good candidates for take-back regulations.

However, take-back requirements may not be cost-effective for all products. Requiring manufacturers of many nondurable goods to take back and recycle their products could simply impose additional costs without clear corresponding environmental benefits. For instance, it would probably not be efficient to collect and recycle potato chip bags; doing so would be likely to cause more pollution from transporting the bags to a recycling facility than would result from landfilling or incinerating

them. And of course, take-back schemes could not be applied to products that are consumed or dissipated during their use.

Take-back requirements have several other limitations. In effect, they impose a predetermined solution (recycling) to the problem of solid waste. They elevate the solid waste aspects of the product above other environmental and performance attributes that may be relevant. If there are design tradeoffs between recyclability and waste prevention, or recyclability and energy efficiency, design decisions may be biased in favor of recyclability, to the detriment of the environment.

Manufacturer take-back programs appear to be moving forward in Europe without any serious attempt at cost-benefit analysis.[144] **OTA suggests that while take-back schemes may be a good option for some products, further research on the costs and benefits for a range of products is needed before they are implemented in the United States (see the discussion of policy research needs above).** These studies should consider the relative merits of market-based incentives such as deposit-refund systems or tradable recycling credit programs as alternatives to take-back regulations.

COORDINATION AND HARMONIZATION

The final area where Congress has a unique role is in coordination and harmonization of policies affecting green design. Green design involves bringing together two policy objectives (industrial competitiveness and environmental protection) that in the past have been seen as separate or even conflicting. It is not surprising, then, that the Federal Government is poorly organized to take advantage of opportunities such as green design. For example, EPA is organized around regulatory responsibilities for protecting air, water, and land; it does not address industrial competitiveness in a natural way, and its technical expertise in design and manufacturing is minimal. The Department of Commerce, on the other hand, is concerned with the competitiveness of industrial sectors, but has little environmental expertise. DOE's national laboratories have a wide range

---

[142] Frances Cairncross, "How Europe's Companies Reposition to Recycle," *Harvard Business Review*, March-April 1992, p. 34.

[143] Several States are enacting take-back laws, such as New Jersey's take-back requirement for rechargeable nickel-cadmium batteries.

[144] "Environmentalism Runs Riot," op. cit., footnote 60.

of technical capabilities that could be brought to bear on improving design for energy efficiency and solid waste recycling processes, but environmental quality has not traditionally been a part of DOE's mission.

Throughout this report, a number of areas have been cited where green design could benefit from a stronger, more coherent Federal approach:

- *Coordinating research.* Projects related to green design are underway in several agency offices (e.g., EPA's Office of Research and Development, Office of Solid Waste, and Office of Pollution Prevention and Toxics; DOE's Office of Industrial Technology; and the National Science Foundation's Engineering Research Centers, see table 6-5), but OTA found that the efforts sponsored by different offices and agencies have often been undertaken independently with little or no coordination among them.
- *Promoting system-oriented design solutions.* Taking advantage of the opportunities for system-oriented green design requires that the economic performance and environmental impact of industries or sectors be viewed in an integrated way. Individual companies have little incentive to promote an overall greener vision of their sector. A greener transportation sector, for example, may involve not only improved vehicle fuel efficiency, but better management of materials used in automotive, rail, and aviation applications, as well as changes in urban design. A coordinated, interagency perspective could spur a more holistic analysis of total sectoral issues, through forums, grant programs, etc.
- *Harmonizing State and Federal environmental product policies.* In the absence of Federal guidance, State and local governments have passed a diverse array of laws affecting the

environmental attributes of products (table 6-2). Industry objects to the prospect of having to comply with a different environmental regime in each State or county, arguing that this is inefficient and inhibits interstate commerce.[145] Environmentalists generally defend the right of each local community to set environmental standards as it sees fit. An interagency forum for discussion and policy development could help define the circumstances under which Federal standards preempting State and local environmental laws may be justified, and where they are not.
- *Coordinating policy development on international aspects of the environment, technology, and trade.* At present, responsibility for development of U.S. policy in these areas is not clearly defined, and each Federal agency has its own agenda.[146]

New Institutions for Environmental Technologies

In Japan, the Ministry of International Trade and Industry (MITI), which has responsibility for both trade and competitiveness, is also involved in implementing Japan's new recycling law. MITI's involvement is expected to be a strong inducement for companies to comply in a timely way.[147] In 1992, a new MITI-run laboratory, the Research Institute for Innovative Technology (RITE) was launched to promote new technologies for improving environmental quality.[148][149] In the United States, however, there is no comparable institution that can address trade, competitiveness, and the environment in a coherent way.

Recently, several proposals have been made to establish a new institutional focus within the Federal Government for integrating environmental and tech-

[145] For example, the Chemical Specialties Manufacturers Association has filed suit in California alleging that California's labeling requirements under Proposition 65 should be preempted by Federal precautionary labeling requirements of the Federal Insecticide, Fungicide, and Rodenticide Act and the Federal Hazardous Substances Act.

[146] William A. Nitze, "Improving U.S. Interagency Coordination of International Environmental Policy Development," *Environment*, vol. 33, No. 4, May 1991, p. 10.

[147] Environmental Resources Limited, op. cit., footnote 136.

[148] Jacob M. Schlesinger, "Thinking Green: In Japan, Environment Means an Opportunity for New Technologies," *Wall Street Journal*, June 3, 1992, p. A1.

[149] RITE's research objectives include development of biodegradable plastics, bioproduction of hydrogen fuels, new metal recovery methods, and new carbon dioxide fixation processes.

nological concerns.[150] These include creating a new Office of National Environmental Technologies within EPA, an independent National Environmental Technologies Agency, a National Institutes of the Environment (analogous to the National Institutes of Health), and a National Environmental Technologies Laboratory within DOE's national laboratory system.[151]

A new institutional focus within the Federal Government for environmental technology could help coordinate Federal efforts to promote various aspects of green design, and provide a home for promising new fields of research such as industrial ecology (see chapter 4), that do not fit readily within any agency's mission. However, OTA does not foresee that a separate institution dedicated exclusively to green design would be appropriate. By its nature, green design is problem-oriented: the appropriate design choices depend on the specific environmental problems to be addressed, and on the particular requirements of various products and industries. For example, packaging designers, auto designers, pesticide formulators, and architects have different information requirements, and operate under different constraints. These would be difficult to address through a single, generic institution.

Interagency Groups

Interagency task forces and committees also provide a mechanism for improving Federal coordination in areas such as environmental policy where no single agency has jurisdiction. In recent years, several interagency groups have been formed to address environmental concerns,[152] for example: the Council on Federal Recycling and Procurement Policy (created in October 1991 to oversee agency recycling actions); the Federal Interagency Task Force on Environmental Labeling (EPA, FTC, and the U.S. Office of Consumer Affairs); the Ad Hoc Committee on Risk Assessment (established in 1990

to harmonize risk assessment approaches among Federal agencies); the Interagency Committee on Environmental Trends (ICET was reactivated by the Council on Environmental Quality in 1991 to coordinate the environmental information activities of various Federal agencies); and the Interagency Task Force on Trade and Environment (led by the Office of the U.S. Trade Representative (USTR)). In 1990, the White House established a subcabinet-level Environmental Policy Review Group under the Domestic Policy Council to review domestic policy issues and improve coordination.[153]

Some relevant interagency collaborations are also being formed on an ad hoc basis. For instance, EPA is working with the Department of Agriculture to promote waste prevention in agricultural chemical use. EPA, DOE, and DOC are collaborating in a joint grant program with States to fund research on reducing the environmental impacts of industrial processes.[154]

Congress could establish a permanent cabinet-level council charged with the responsibility of ensuring that environmental concerns are integrated into all Federal policies. This might take the form of an expanded Council on Environmental Quality, or a new Environmental Policy Council with its own permanent staff.[155] [156] To be taken seriously, though, such a council would have to enjoy the full support of the President.

Alternatively, Congress can use its oversight powers to ensure that the activities of existing interagency groups are consistent with green design. For example, it can ensure that: waste prevention is incorporated into procurement initiatives developed by the Council on Federal Recycling and Procurement Policy; mechanisms for coordinating Federal data collection on toxic materials flows are considered by the Interagency Committee on Environmental Trends; and that the USTR-led Task Force on Trade and the Environment has adequate

---

[150] See "Senate, House Members Craft Bills To Push Federal 'Green' Technology Policy," *Inside EPA*, July 3, 1992, p. 17; Helen Gavaghan, "Green Research Gains Ground in America," *New Scientist*, Apr. 18, 1992, p. 8; Braden Allenby, AT&T, "Why We Need a National Environmental Technology Laboratory (And How To Make One)," unpublished draft.

[151] At this writing, these and other proposals were under review by the Task Force on Environmental Research and Development of the Carnegie Commission on Science, Technology, and Government, and National Academy of Sciences' Committee on Environmental Research.

[152] Council on Environmental Quality, op. cit., footnote 8.

[153] William A. Nitze, op. cit., footnote 146, p. 32.

[154] The program is called National Industrial Competitiveness through Efficiency: Energy, Environment, Economics (NICE³).

[155] Alvin L. Alm, "A Need for New Approaches," *EPA Journal*, May/June 1992, p. 7.

[156] U.S. Environmental Protection Agency, Science Advisory Board, op. cit., footnote 98, appendix C, p. 56.

policy guidance in international negotiations on environmental product policies.[157]

Technology With a Green Lining

Regardless of whether Congress creates any new environmental technology institutions, OTA believes it makes sense to integrate environmental concerns more thoroughly into each agency's ongoing programs. One recent study has developed a list of "environmentally critical technologies."[158] But ideally, there should be an environmental component to each of the "critical" technologies on the lists already assembled by the Office of Science and Technology Policy, the Department of Commerce, and the Department of Defense. The goals of waste prevention and better materials management could be integrated thoroughly into NIST's Advanced Technology Program,[159] the recently announced Manufacturing Technology Initiative,[160] and the Advanced Materials and Processing Program.[161] **Congress can use its oversight powers to ensure that both new and existing technology development programs have an environmental dimension.**

In the end, the institutional details are less important than a recognition on the part of Congress and the Administration that Federal leadership is needed to take advantage of opportunities like green design that do not fall neatly within the mission of any single agency.

A STARTING POINT

Many of the options discussed above would not immediately affect the way products are designed. Research to define environmental risks and understand life-cycle materials flows will take time. Changes in the curricula of design and engineering schools will affect the next generation of designers. And changes in the tax code to internalize the environmental costs of materials and energy use and product disposal do not appear to be on the political horizon, particularly in an era of concern about economic growth and U.S. industrial competitiveness.

OTA believes that such long-term changes are essential if the United States is to be a world leader in green design. But a shorter-term strategy is also important to ensure that existing momentum is not lost. The following is a package of options Congress might consider that could be implemented relatively quickly, and would not be very expensive:

- Require all Federal agencies to conduct a thorough review of their regulations and procurement policies (including military specifications) that may discourage waste prevention and better materials management, and make recommendations for changes. These changes would be consistent with the Federal Recycling and Procurement Policy (Executive Order 12780) and would not require any new legislative authority.

- Provide funding to EPA to expand the Pollution Prevention Information Exchange System to include all Federal and State activities relevant to green design in a single place. An electronic network would stimulate cross-fertilization of current projects and help eliminate duplication of effort.

- For products with significant environmental impacts (e.g., autos, paper, pesticides, etc.), provide funding through the appropriate agencies for intensive workshops that would bring together professionals associated with various phases of a product's life cycle (e.g., designers, suppliers, manufacturers, distributors, consumer advocates, and waste management providers) to discuss opportunities for coordinated action for waste prevention and better materials management.

---

[157] U.S. Congress, Office of Technology Assessment, *Trade and Environment: Conflicts and Opportunities*, OTA-BP-ITE-94 (Washington, DC: U.S. Government Printing Office, May 1992).

[158] George R. Heaton, Jr. et al., "Backs to the Future: U.S. Government Policy Toward Environmentally Critical Technology," World Resources Institute, Washington, DC, June 1992.

[159] The ATP program is primarily oriented toward enhancing U.S. competitiveness. Of the 27 ATP grants awarded in 1992, several are indirectly related to environmental concerns, though only one is directly related (a project on plastics recycling).

[160] "Technology Initiative Initiated," *Science*, vol. 255, Mar. 13, 1992, p. 1350.

[161] A number of environment-related projects are proposed in Advanced Materials and Processing: the Fiscal Year 1993 Program, op. cit., footnote 51.

- Provide funding for a national green design competition and establish a prestigious National Green Design Award similar to the Malcolm Baldridge National Quality Award.[162] [163]

A design competition and national award would generate new ideas for designers across the country, and give consumers a better sense of the possibilities.

---

[162] U.S. Congress, Office of Technology Assessment, *Facing America's Trash*, op. cit., footnote 109, p. 24.

[163] The Malcolm Baldridge Award does include criteria such as waste prevention, but the environment is not a central focus. See ''Malcolm Baldridge National Quality Award, 1992 Award Criteria,'' available from the U.S. Department of Commerce, Technology Administration, Washington, DC, 1992.

Chapter 7

Guiding Principles for Policy Development

Contents

Guiding Principles for Policy Development

As discussed in the preceding chapter, the flow of materials and products through the economy gives rise to environmental impacts that are not adequately accounted for under current environmental policies. Table 6-6 lists a variety of regulatory and market-based options that have been proposed to address these impacts. These involve government intervention at various stages of the product life cycle, ranging from taxes on production of virgin materials to waste disposal fees. All of these options could have an impact on the product design process.

What criteria can policymakers use to evaluate these options? The Office of Technology Assessment (OTA) suggests three guiding principles that can help to shape environmental policies that encourage, rather than inhibit, green design:

- Identify the root problem and define it clearly.
- Give designers the maximum flexibility that is consistent with solving the problem.
- Encourage a systems approach to green design.

These principles are developed further below. The chapter concludes with a broader perspective on the significance of green design for U.S. competitiveness and environmental quality.

GENERAL PRINCIPLES

Principle 1: Identify the Root Problem and Define It Clearly

One of the biggest challenges in developing a policy is clearly defining the environmental problem to be addressed. Often, products and waste streams have multiple environmental impacts that cannot be easily disentangled. For example, policymakers may be concerned with the quantity of a particular waste stream, its toxicity, or its persistence in the environment. Policies aimed at solving a problem at one stage of the life cycle may have unintended negative effects at another stage: for example, requiring automobiles to be made from currently recyclable materials could adversely affect their fuel efficiency. Inevitably, tradeoffs and value judgments must be made to determine which environmental impacts are the most important.

Despite the difficulty, the discipline of defining the problem clearly is critical to defining an appropriate policy response. In the absence of a clearly defined problem, it becomes easy to confuse means and ends. In the Resource Conservation and Recovery Act (RCRA) reauthorization debate, for example, the problem is often framed in terms of the large quantity of municipal solid waste being generated. But a solution often put forward is to mandate higher recycling rates—as if the problem was that recycling rates are too low. The figure of merit for measuring progress then becomes higher recycling rates, instead of less waste generated.

This approach misses the point that recycling is only one of several means to reduce the quantity of solid waste destined for disposal. Perversely, an exclusive emphasis on recycling could even lead to more waste being generated, especially if such emphasis discourages designs featuring waste prevention. If the objective is to reduce the amount of solid waste generated, municipal solid waste policies and government procurement programs should make allowances for product designs that feature waste prevention.

Without a clearly defined problem, there is a tendency to focus on the most visible environmental issues, rather than those that are the most important. Recent examples include proposals to ban heavy metals from packaging[1] (despite the fact that packaging is a minor source of heavy metals in landfills and incinerators) and proposals to regulate municipal trash (while the much larger problem of industrial solid waste has not been addressed).[2]

A clearly defined problem can also help to set priorities. For example, although the dissipation of toxic materials in the global environment is a growing problem, not all toxic chemicals and products are of equal concern. Treating them as equal can divert attention and resources from truly

---

[1] Model legislation developed by the Coalition of Northeastern Governors' Source Reduction Task Force has been passed in 10 States.

[2] U.S. Congress, Office of Technology Assessment, *Managing Industrial Solid Wastes From Manufacturing, Mining, Oil and Gas Production, and Utility Coal Combustion*, OTA-BP-O-82 (Washington, DC: U.S. Government Printing Office, February 1992).

high-risk chemicals and waste streams.[3] Similarly, comprehensive reporting requirements on industrial use of all toxic materials are not necessarily cost effective. Proposals to require companies to report on their use of hundreds of additional chemicals, without distinguishing those chemicals that are of greatest concern, could generate a massive paperwork burden without significant environmental benefits.[4]

In reauthorizing RCRA and other environmental legislation, Congress has an opportunity to refocus attention and resources on the key problems associated with current materials flows. If it frames the objective in terms of reducing the generation of wastes, especially those that pose the greatest risks, it will encourage the design of products that use resources efficiently and waste management programs that are cost-effective. If, on the other hand, it frames the objective in terms of increased recycling rates, and if it fails to distinguish high-risk waste streams from low-risk waste streams, it may encourage less efficient product designs and less efficient waste management programs.

Principle 2: Give Designers the Maximum Flexibility That Is Consistent With Solving the Problem

Materials technology options are proliferating rapidly, and product impacts on the environment are multidimensional. This suggests that policies should be crafted to give designers as much flexibility as possible, within a framework that protects human health and the environment. This can be accomplished by several means, as discussed below.

Voluntary Agreements With Industry

Perhaps the greatest flexibility can be achieved through negotiated voluntary agreements between government and industry. Such agreements tend to be easier and faster to implement than legislation and regulations, and may be attractive to industry because it has more control over the targets and timetables. Several countries in Europe are relying more heavily on voluntary negotiations with industry to achieve waste reduction goals, especially the Netherlands, Germany, Sweden, and Denmark (see chapter 5). In the case of the German proposals requiring manufacturers to take back packaging and automobiles, industry has been given the opportunity to develop its own system for collecting and recycling the products before more heavy-handed regulations or mandatory deposit-refund systems go into effect.

In the United States, the Environmental Protection Agency (EPA) is also moving in the direction of voluntary programs; examples include the 33/50 Program and the Green Lights Program.[5] An example at the State level is Massachusetts' Toxics Use Reduction Act, which emerged from negotiations involving government, industry, and public interest groups. Under this law, companies are required to develop facility plans with self-set goals to reduce the use—not just the release—of toxic chemicals.[6]

It must be said that in most cases, such voluntary agreements are driven by public and political pressure, the threat of tough new laws and regulations, or imminent enforcement actions. In general, credible mechanisms for monitoring and enforcement of voluntary agreements are still being developed.

Flexible Regulations

Regulations affecting product design can be crafted with built-in flexibility. For instance, an important policy objective is to find ways to credit waste prevention in recycling legislation and in government procurement programs for recycled goods. One option is to provide alternative criteria for acceptable products; instead of imposing mandatory recycled content requirements for packaging, acceptable packaging could contain a certain percentage recycled content *or* a certain percentage weight reduction, etc.[7] This more flexible approach takes into account the inherently multidimensional

[3] Michael M. Segal, "Spilled Some Salt? Call OSHA," *Wall Street Journal*, July 9, 1991, p. A16.

[4] See, e.g., testimony of the American Petroleum Institute, before the Subcommittee on Transportation and Hazardous Materials, House Committee on Energy and Commerce, Mar. 10, 1992.

[5] U.S. Environmental Protection Agency, "Pollution Prevention Resources and Training Opportunities in 1992," EPA/560/8-92-002, January 1992, p. 84; John S. Hoffman, "Pollution Prevention as a Market-Enhancing Strategy: A Storehouse of Economical and Environmental Opportunities," *Proceedings of the National Academy of Sciences*, vol. 89, February 1992, p. 832.

[6] Ken Geiser, "The Greening of Industry," *Technology Review*, August/September 1991, p. 64.

[7] This approach, originally developed by the Massachusetts Public Interest Research Group (MASSPIRG), has become part of the RCRA reauthorization debate.

nature of green design, and helps to avoid forcing suboptimal design solutions to the solid waste problem.

One difficulty with this approach is that a more flexible regulation may be more expensive to monitor and enforce. For example, waste reduction is difficult to measure. It is likely to be easier to verify that a package contains a certain percentage of recycled content than to verify that it uses a certain percentage less material than a comparable package did 5 years ago. It seems likely that industry will have to bear the burden of demonstrating compliance with flexible regulations if this approach is going to work.

Economic Instruments

Market-based policy instruments such as emissions taxes, tradable emissions permits, or deposit-refund systems will generally provide a more flexible environment for product design than regulations, because designers are free to make choices based on minimizing overall costs. Economic instruments may be used in place of regulations or as a supplement to make their implementation more flexible and cost effective. For example, a regulation requiring recycled content in products may be implemented more cost effectively by establishing a tradable recycling credit scheme to encourage those manufacturers who can incorporate secondary materials most cheaply to do so.[8]

A potential disadvantage of market-based instruments such as tradable recycling credits is the cost of monitoring and enforcement. For example, verifying that a manufacturer has purchased sufficient credits to cover the virgin material content of his or her product may be difficult. Such market-based instruments may work best for a limited number of products of special concern.[9]

Principle 3: Encourage a Systems Approach to Green Design

Designers can control many of the environmental attributes of products, but they have only a limited ability to influence the systems by which products are manufactured, used, and disposed (see chapter 4). For example, a designer can make a product more recyclable by making it easier to disassemble into component parts, but if there is no infrastructure in place to recover and recycle the product, the benefits of the design changes are nullified. Coupling product design with recycling implies the formation of new relationships among materials suppliers, manufacturers, and waste management providers. Often, however, the incentives for changing these interfirm linkages are lacking; companies may also have large capital investments in existing production and distribution networks. Therefore, policy incentives are needed to provide the impetus for change.

Incentives for a Systems Approach

A system-oriented design approach can be encouraged by improving the linkages between design decisions and their environmental consequences. This can be accomplished either directly by regulation, or indirectly through taxes or other economic instruments that internalize environmental costs.

Recycled content regulations or manufacturer take-back requirements are examples of regulatory coupling between two stages of the product life cycle: manufacturing and waste management. These regulations can help make solid waste concerns a key design consideration. For example, the proposal of the German Government to require auto manufacturers to take back and recycle their cars has stimulated designers to rethink the entire ''ecology'' of auto production and disposal (box 4-F). Perhaps to head off similar regulation in the United States, Ford, General Motors, Chrysler, their suppliers, and the auto recycling industry have formed a consortium called the Vehicle Recycling Partnership to address the recycling issue.[10] However, as discussed in chapter 6, such take-back regulations may be more efficient for some products (especially high-value durable goods or products that pose special waste management problems) than others.

Economists have long argued that it is not necessary to close materials cycles through recycling regulations if the prices of goods and services reflect their full social costs.[11] If the economic circle

[8] Michael H. Levin, ''Implementing Pollution Prevention: Incentives and Irrationalities,'' *Journal of Air and Waste Management Association*, vol. 40, No. 9, September 1990, p. 1227.

[9] Organisation for Economic Opportunity and Development, *Environmental Policy: How To Apply Economic Instruments*, Paris, 1991, p. 107.

[10] On March 16, 1992, organizers held the First Vehicle Recycling Partnership Forum in Dearborn, MI.

[11] William D. Nordhaus, ''The Ecology of Markets,'' *Proceedings of the National Academy of Sciences*, vol. 89, February 1992, p. 843.

is closed, they say, the market will sort out the most efficient systems of production and materials management. Notwithstanding these advantages, however, environmental policies in all countries are primarily based on regulations. In part, this is due to the fact that no mechanism exists to establish the "true" value of environmental services. Another reason is that environmental taxes tend to be politically unpopular.

However, there is now renewed interest in the use of economic instruments in environmental policy, if not to replace regulations, at least to supplement them and help make them work more efficiently.[12] There is also interest in the idea of shifting the tax burden from socially desirable activities such as savings and work to undesirable activities such as pollution.[13]

Such a shift could have a dramatic impact on the systems by which products are manufactured, distributed, used, and disposed. For example, a phased-in $100 per ton carbon tax on fuels could not only encourage more efficient use of materials and energy in production systems, but could also transform consumption patterns and raise over $100 billion in government revenues.[14]

GREEN DESIGN IN PERSPECTIVE

How should one view the significance of green design as a competitive and environmental strategy? As a competitive strategy, green design can help manufacturers generate less waste and reduce production costs at the same time.[15] As waste disposal costs and regulatory compliance costs go up, the environmental attributes of products will necessarily become more important to consumers and investors. Europe and Japan are already moving aggressively to integrate "clean" technology and products into their industrial strategies for future competitive-

ness,[16] and international trade will increasingly be influenced by environmental concerns.[17] All of these trends suggest that having an environmental dimension to one's design capabilities will be an important competitive asset in the future.

As an environmental strategy, green product design offers a new way of addressing environmental problems. By recasting pollution concerns as product design challenges, and particularly by encouraging designers to think more broadly about production and consumption systems, policymakers can address environmental problems in ways that would not have been apparent from a narrow focus on waste streams alone.

However, the significance of green design for overall environmental quality is harder to assess. Individual designers will no doubt find many opportunities to reduce wastes and increase production efficiencies. But designers operate within the constraints of available manufacturing process technologies, waste management infrastructure, and government policies on resource use and economic development. For instance, a housing development built on an environmentally sensitive wetland can hardly be considered green, even if the units are energy efficient and made with recycled materials. The potential of green design to address environmental problems is therefore contingent on broader environmental policies.

As discussed in Chapter 3, U.S. environmental policies are currently based on the environmental protection paradigm, being concerned mainly with ameliorating the effects of human activities on the environment.[18] Generally, this has meant end-of-pipe pollution controls and after-the-fact cleanups where allowable pollution limits have been exceeded. Recently, however, the emphasis has begun to shift toward waste prevention strategies.

[12] Robert W. Hahn and Robert N. Stavins, "Incentive-Based Environmental Regulations: A New Era From an Old Idea?" Energy and Environmental Policy Center, John F. Kennedy School of Government, Harvard University, E-90-13, Cambridge, MA, August 1990.

[13] See, e.g., George Heaton et al., *Transforming Technology: An Agenda for Environmentally Sustainable Growth in the 21st Century* (Washington, DC: World Resources Institute, April 1991).

[14] U.S. Congress, Congressional Budget Office, *Carbon Charges as a Response to Global Warming: The Effects of Taxing Fossil Fuels* (Washington, DC: U.S. Government Printing Office, August 1990).

[15] Bruce Smart (ed.), *Beyond Compliance: A New Industry View of the Environment* (Washington, DC: World Resources Institute, April 1992).

[16] See, e.g., Jacob M. Schlesinger, "Thinking Green: In Japan, Environment Means an Opportunity for New Technologies," *Wall Street Journal*, June 3, 1992, p. A1.

[17] U.S. Congress, Office of Technology Assessment, *Trade and Environment: Conflicts and Opportunities*, OTA-BP-ITE-94 (Washington, DC: U.S. Government Printing Office, May 1992).

[18] Michael E. Colby, *Environmental Management in Development: The Evolution of Paradigms* (Washington, DC: The World Bank, December 1990).

In the context of the environmental protection paradigm, green design can be a useful tool to increase industrial efficiency and to complement waste prevention strategies. But critics of this paradigm argue that simply increasing the efficiency of materials and energy use and reducing pollution rates may not be enough to ensure the future survival of the ecological systems upon which the economy and human life depend. It is quite possible to destroy the environment while continuing to become more efficient. Progress must be measured, these critics say, not by marginal reductions in pollution based on last year's levels, but by cumulative damage to ecological systems and their general sustainability.

Over the years, several national commissions and studies have examined the appropriate Federal role in managing resource use and materials flows.[19] The focus of these studies has gradually shifted from a concern with ensuring the availability of future resources for industry to a concern with managing materials use under an increasing number of constraints, including environmental constraints.[20] Nevertheless, current U.S. environmental policies are not explicitly concerned with managing the physical flows of energy and materials through the economy in ways that are ecologically ''sustainable.'' The Federal Government has been reluctant to address issues of materials management directly, preferring to leave these decisions to the States and private industry.[21] Despite its title, for instance, the Resource Conservation and Recovery Act is primarily concerned with regulating the disposal of hazardous wastes, not conserving or recovering resources.

Both the resource management and the eco-development paradigms are explicitly concerned with conservation and sustainability of materials use. In the case of the resource management paradigm, this is accomplished through closing the economic loop by internalization of environmental costs; in the eco-development paradigm, the emphasis is on substituting renewable for nonrenewable resources, reducing use of toxic chemicals, and closing materials loops through recycling of nonrenewable resources.[22] Under either of these paradigms, green design is not simply a useful tool, but an essential strategy for resource conservation and sustainable materials management.

In a world where population growth and economic growth put increasing pressures on natural resources and ecosystems, the dominant paradigm upon which environmental policies are based can be expected to evolve from environmental protection toward resource management and eco-development.[23] Policies in several countries, especially Germany and the Netherlands, are already beginning to reflect this shift (see chapter 5). As this evolution occurs, the importance of green design can be expected to grow. Therefore, policymakers should strive to make green product design an integral part of strategies to improve competitiveness and environmental quality.

OTA's investigations suggest that simply providing better information to designers and consumers about the environmental impacts of products and waste streams will not be enough. To move ahead, policies must provide a closer coupling between design decisions and their environmental consequences. The challenge to policymakers is to choose a mix of regulatory and economic instruments that target the right problems and give designers the flexibility to find innovative, environmentally elegant solutions.

---

[19] For a review, see Resource Conservation Committee, ''Choices for Conservation,'' *Final Report to the President and Congress,* SW-779, July 1979, p. 33.

[20] Ibid.

[21] OTA has previously discussed the value of a national materials management policy in the context of municipal solid waste. See U.S. Congress, Office of Technology Assessment, *Facing America's Trash: What Next for Municipal Solid Waste?* (Washington, DC: U.S. Government Printing Office, October 1989), p. 6.

[22] Michael E. Colby, op. cit., footnote 18.

[23] Ibid.

Superintendent of Documents **Publications** Order Form

Order Processing Code:
* **5335**

Charge your order.
It's Easy!

To fax your orders **(202) 512–2250**

☐ **YES**, please send me the following:

_____ copies of *Green Products by Design: Choices for a Cleaner Environment (128 pages)*
S/N 052-003-01303-7 at $6.50 each.

The total cost of my order is $_____ . International customers please add 25%. Prices include regular domestic postage and handling and are subject to change.

_____ (Please type or print)
(Company or Personal Name)

_____
(Additional address/attention line)

_____
(Street address)

_____
(City, State, ZIP Code)

_____
(Daytime phone including area code)

_____
(Purchase Order No.)

May we make your name/address available to other mailers?   **YES   NO**   ☐ ☐

Please Choose Method of Payment:

☐ Check Payable to the Superintendent of Documents

☐ GPO Deposit Account ☐☐☐☐☐☐–☐

☐ VISA or MasterCard Account

☐☐☐☐☐☐☐☐☐☐☐☐☐☐☐☐☐☐☐☐

☐☐☐ (Credit card expiration date)   *Thank you for your order!*

_____   9/92
(Authorizing Signature)

Mail To: New Orders, Superintendent of Documents
P.O. Box 371954, Pittsburgh, PA 15250–7954

P3

ISBN 0-16-038066-9

90000

9 780160 380662